James Monteith

Physical Geography of Western Tibet

James Monteith

Physical Geography of Western Tibet

ISBN/EAN: 9783742853189

Manufactured in Europe, USA, Canada, Australia, Japa

Cover: Foto ©Klaus-Uwe Gerhardt /pixelio.de

Manufactured and distributed by brebook publishing software
(www.brebook.com)

James Monteith

Physical Geography of Western Tibet

CHARACTER OF THE WORK.

Its Style.—In the preparation of this portion of the work, the author has sought to present the subject of Geography *as a Science;* and, at the same time, in a style calculated to attract and interest the pupil. Avoiding the use of all technical terms that would perplex the young learner, he has endeavored to explain its leading principles by means of familiar language and comparisons, and of suggestive illustrations, according to the *Object System* of instruction. For example, on page 19, the theory of volcanic action is explained by reference to a cake which is burst open at the top, the heat of the oven representing the heat of the earth's interior. Again; on page 27, boiling springs, such as the geysers, are illustrated by means of a tea-kettle.

The Text is divided into Short Paragraphs so constructed that the commencement of each appears in prominent type and readily suggests the subject and the questions.

It Teaches;—That the earth was formed to be the temporary dwelling-place of mankind; and to that end were created the land, with its mountains and plains; the water, with its mighty ocean and its running brooks; besides air, light, heat, plants, and living creatures;

That all the objects which we behold, whether organic or inorganic, whether on the surface or below the surface, with all the various phenomena of the earth, exert important influences upon each other and work together in harmony for the well-being of the human race.

Effect upon the Mind of the Pupil.—Throughout the work the aim has been not only to impart valuable information, but especially to cultivate the learner's powers of observation and reasoning; and, as he views the wonders, the beauty, and the perfection of Nature's works, his thoughts are thereby directed to the Creator, by whom all things were made and adapted to the development of human life and happiness.

The Index and General Review serve both as a Reference and as a system of General Exercises.

CONTENTS.

SECTION I.
INTRODUCTORY—THE EARTH THE DWELLING-PLACE OF MANKIND.................... 5

SECTION II.
THE CREATION OF THE EARTH—ITS CHANGES AND GRADUAL DEVELOPMENT—FORMATION OF SOIL—COMMENCEMENT OF VEGETABLE AND ANIMAL LIFE.......... 6

SECTION III.
THE CRUST OF THE EARTH—INTERNAL HEAT—STRATA............................. 8

SECTION IV.
THE FORM OF THE EARTH—HOROSCOPE. WATER, LAND, AIR, LIGHT, AND HEAT NECESSARY TO MAN'S EXISTENCE—THE HARMONY OF NATURE'S LAWS............. 9

SECTION V.
THE FORMATION OF CONTINENTS—UPHEAVAL AND SUBMERGENCE—THE WISDOM OF GOD'S PLAN MANIFESTED—MOUNTAIN SYSTEMS—THE LAND AND WATER HEMISPHERES—INLETS, RIVERS, ETC., ADVANCE THE CIVILIZATION OF MAN........ 10

SECTION VI.
MOUNTAINS AND PLATEAUS; THEIR ORIGIN, PLACES, AND USES—GLACIERS—MOUNTAIN PASSES.................................... 14

SECTION VII.
VOLCANOES AND EARTHQUAKES; THEIR ORIGIN AND EFFECTS.......................... 19

SECTION VIII.
PLAINS AND VALLEYS; THEIR DISTRIBUTION; HOW THEIR SOIL IS ENRICHED.... 20

SECTION IX.
DESERTS AND OASES; THEIR DISTRIBUTION; CAUSES OF THEIR FORMATION............ 21

SECTION X.
THE OCEAN; ITS EXTENT AND DIVISIONS; ITS DEPTH AND BED; ITS SALTNESS...... 22

SECTION XI.
OCEANIC CURRENTS; THE THEORY OF THEIR MOVEMENTS; THEIR IMPORTANT INFLUENCES AND BENEFITS..................... 23

SECTION XII.
EVAPORATION—SPRINGS AND WELLS; THEORY OF THEIR FORMATION—THE GEYSERS. 26

SECTION XIII.
RIVERS; THEIR ORIGIN, POWERS, AND IMPORTANCE TO MAN................... 28

SECTION XIV.
LAKES; THEIR FORMATION, ELEVATION, AND DEPTH................................ 31

SECTION XV.
THE ATMOSPHERE—THE WINDS—LAND AND SEA BREEZES........................ 32

SECTION XVI.
VAPOR—CLOUDS—DISTRIBUTION OF RAIN.... 35

SECTION XVII.
CLIMATE; ITS DEPENDENCE UPON OCEANIC CURRENTS AND WINDS; ITS INFLUENCE UPON VEGETATION AND MAN—ISOTHERMS AND CLIMATIC ZONES—THE CLIMATE OF ELEVATED REGIONS...................... 36

SECTION XVIII.
VEGETATION; ITS GROWTH AND USES; ITS DISTRIBUTION—THE FORMATION AND DISTRIBUTION OF COAL FIELDS............. 40

SECTION XIX.
ANIMALS; THEIR CREATION, GRADUAL DEVELOPMENT, AND USES; THEIR ADAPTATION TO CLIMATES AND OTHER CONDITIONS.................................. 42

SECTION XX.
MANKIND—THE RACES—THE INFLUENCES OF CLIMATE, FOOD, AND MEANS OF INTER-COMMUNICATION, UPON INDIVIDUALS AND NATIONS.................................. 44

INTRODUCTORY.

1. The robin builds her nest in the tree for the *Purpose* of there depositing her eggs, and of bringing forth and protecting her young.

2. For the *Purpose* of protection and comfort men build houses, found cities, and establish governments. *Purpose, therefore, leads to Design and Action.*

3. When you look at a beautiful house, and observe the peculiar fitness of the various parts to each other, you are certain that it was made for the security and enjoyment of the family within; and that the workmen shaped and placed the materials under the direction of an intelligent architect, who *Formed the Plan before the Work was commenced.*

4. So, when you look abroad, you see a beautiful world, which was made for the enjoyment and benefit of the whole human family.

5. *Man could not exist without Food;* therefore the earth yields her manifold productions of grain, fruit, and vegetables, while animals, birds, and fish, also, are given for his nourishment and use. *Neither could he live without Drink;* so the earth is abundantly supplied with refreshing springs. *For Clothing* he goes to the cotton plant, the sheep, and the silkworm; from the forests and the ground he obtains all the materials for building purposes.

6. *Animal Life receives its Sustenance* from plants; *Plants receive theirs* from the soil and moisture; *Soil proceeded* originally from the hard rock; *Moisture and Clouds,* from the ocean.

7. The earth has its continents and oceans, its mountains and plains, its rocks, rains, snows, springs, and streams. All work harmoniously for the welfare and happiness of mankind.

8. You may conclude, then, that the whole earth, of which all these things are but parts, was made for a *Great Purpose,* by a Being of infinite wisdom, goodness, and power, according to a design formed before the beginning of the world; and this purpose was to provide an *Abode for Man, whose Delight would be to praise, honor, and serve Him.*

Section II.

CREATION OF THE EARTH.

The Earth's Surface covered with Water.

1. The *Growth of a Plant* progresses slowly and systematically; from the seed comes a stem, then leaves, blossoms, and fruit. So was the process by which the world was made from chaos,—slow, gradual, and in accordance with the provisions of a well-ordered plan established by Divine wisdom.

2. The *Earth's Formation from Chaos* may be illustrated by an egg, whose fluid substances, by a certain application of heat, and in a certain time, are changed into a beautiful, living bird.

3. "*In the Beginning*, God created the heaven and the earth." In time the earth received its globular shape, and consisted of a heated, earthy matter in a fluid form, the outside of which, becoming cool and hard, formed a kind of crust around the mass. Entirely surrounding this crust was water, and surrounding the water was the atmosphere, containing dark, heavy clouds.

4. The *Rain formed no Springs*, watered no fields. It fell only upon the salt ocean, for the whole outer side of the earth's crust constituted the bed of the ocean.

" And God said, Let the Waters under the Heaven be gathered together, and let the Dry Land appear."

5. *By Convulsions within the Earth*, parts of the crust were forced upward through the water, and became dry land.

6. The *Land first Raised* consisted only of masses of hard rock, on which no tree or plant could grow.

7. *There was no Soil* until the rock was broken and pulverized by the action of the waves, air, rain, heat, and cold.

8. From the grinding together of fragments of the rock, came stones, pebbles, gravel, and sand.

9. "And God called the dry land Earth, and the gathering together of the waters called He Seas."

"And the Earth brought forth Grass, and Herb yielding Seed after his kind, and the Tree yielding Fruit."

10. The *Violent Agitation of the Earth's Interior* greatly disturbed the bed of the ocean, causing the depression of some parts and the elevation of others; in the former, the sea became deeper, and in the latter, more shallow.

11. *Portions of the Ocean's Bed* were in this way brought up to the surface, then above it; and, covered with the pulverized or disintegrated rock which had long been settling upon them, these tracts of land, in time, supported trees and plants which received their nourishment from both the soil and the atmosphere.

" And God created great Whales, and every Living Creature that moveth, which the Waters brought forth abundantly, after their kind, and every Winged Fowl after his kind."

"And God made the Beast of the Earth after his kind, and Cattle after their kind, and every thing that creepeth upon the Earth after his kind."

12. The *Various Species of Animals* which have lived upon the earth were not all created at once.

13. The *Lower Orders* came first; and, as centuries rolled on, other and superior classes of animals came successively into existence.

14. *Insects, Fish, and Reptiles* were created before the horse or the ox; and all species of animals were created before Man.

15. *With Plants, also, this was the case.* The first vegetation consisted of sea-weeds; then, with the improvement of the soil, new and superior varieties of plants and trees appeared.

16. *These facts have been ascertained* from investigations below the earth's surface, where the forms or remains of plants and animals, which lived in successive periods, are found in the order of their creation; those created first being furthest below the surface.

17. *We see the Law of Gradual Development* exemplified in the growth of the trees and living creatures; geologists observe it, also, in the rocks and sands of the earth.

"And God said, Let us make Man in our image, after our likeness; and let them have dominion over the Fish of the sea, and over the Fowl of the Air, and over the Cattle, and over all the Earth, and over every Creeping Thing that creepeth upon the Earth."
"So God created Man in his own Image."

18. The general order of Creation was as follows:
(1.) CHAOS.
(2.) MELTED MATTER in the form of a globe.
(3.) The GLOBE composed of melted matter having a crust which was entirely surrounded by water.
(4.) PARTS OF THE CRUST upheaved through the sea, forming dry land.
(5.) PULVERIZED ROCK; forming soil.
(6.) LAND ALTERNATELY UPHEAVED AND SUBMERGED.
(7.) VEGETATION.
(8.) ANIMAL LIFE.
(9.) MAN.

19. The *Observing Pupil has now Learned* two important facts; first, that God made the world, with all it contains, not at once, but step by step, on a wise and definite plan; second, that He made it for the use of man.

20. *For the Life and Happiness of Mankind* there are provided, not only the objects and creatures mentioned in the beginning of Genesis, but also numberless features and phenomena of the earth, such as its atmosphere, climates, currents, rain, mountains, plains, and productions.

21. The *Science of Geography* properly embraces an investigation into the laws which control the conditions, changes, and phenomena in nature, as affecting the life and condition of mankind.

22. Although the various departments of Geographical Science will be presented in this work in a classified form, yet it is highly important that the learner keep constantly in mind their dependence and influence upon each other; this renders repetition, to some extent, essential.

23. When considering the position and height of a chain of mountains, the course of the winds, or of an ocean current, he should observe the influences exerted by each upon climate, vegetation, and the pursuits of man in different regions.

24. These *Differences or Contrasts* furnish each section with its own characteristic productions, and lead men to establish a system of trade or commerce between the nations of the earth, thus increasing their industry and wealth, furnishing incentives for exploration, and securing the civilization and enlightenment of the race.

25. The *Pupil should know*, not only that the Gulf Stream has a north-easterly direction, but also that its warmth tempers the climate of the greater part of Europe, and sheds its genial influence upon the atmosphere, productions, and inhabitants of that Grand Division. He should observe that the highest mountains are in the hot regions of the earth, where their lofty peaks, continually wrapped in snow, are faithful refrigerators, reducing the temperature of the air on the heated plains below.

26. The text, generally, is written without set questions, leaving the teacher to frame or vary them as he may wish. Interrogations, however, are made which can be answered not directly from the text, but from the illustrations, or by inference on the part of the learner.

27. This plan cannot fail to lead youthful minds to habits of observation and reasoning, and to direct their thoughts to the

Section III.

THE CRUST OF THE EARTH.

1. The *Crust of the Earth is the result of* the cooling of the melted matter at the surface. It becomes thicker, as ice does, by additions to its under side.

2. Scientific investigations show that the ground is affected by the sun's heat to the depth of about 50 feet; below that, the heat of the earth's interior increases according to the depth.

3. The *Internal Heat* does not extend to the surface of the earth, except on occasions of earthquakes and volcanic eruptions.

A View within the Earth's Crust, by a Mine of Austria, 500 Feet Deep.

4. The average increase of temperature, below where it is affected by heat from the sun, is about one degree for every 50 or 60 feet in depth; accordingly, at the depth of about 60 miles, the heat would be sufficient to melt all known rocks.

5. Geologists have variously estimated *the thickness of the earth's crust* to be from 20 to 200 miles.

6. Man has penetrated the earth to the depth of about one mile.

7. The *Crust*, if 20 miles in thickness, bears the same proportion to the whole earth that an egg-shell does to the egg.

8. The *Height of the Highest Mountains* in the world is about 5 miles, yet the distance from the level of the sea to the center of the earth is 800 times greater than that.

9. The *Material* of which the earth's crust is composed is termed Rock, whether it be hard and compact, or soft and loose; it is constantly undergoing change, owing, chiefly, to the agency of air, water, and heat.

10. *Aqueous Rocks* are those formed by the agency of water. They consist of the sediment which has become hardened in layers or beds, and are called Stratified.

11. *Igneous Rocks* are those formed by the agency of fire. They consist of hard, irregular masses, and are therefore called Unstratified.

A. Stratified Rock; B. Unstratified Rock; C. Melted Matter of the Earth's Interior.

12. *As the Surface*, at an early period, was entirely covered with water, where would you find the Aqueous or Stratified formations?

13. The *Igneous or Unstratified* rock found at the earth's surface has been forced up through the aqueous or stratified formations by volcanic action.

14. *In some Rocks are found* forms of animals and vegetables petrified or hardened like stone, caused, chiefly, by chemical action in nature.

15. *Geologists show* that the greater part of the soil or mold on the earth's surface is composed of what in former ages constituted the bodies of animals, trees, and plants, mixed with mineral substances, all of which settled at the bottom of the water.

16. The petrified forms of animals and plants are called *Fossils*; the strata in which they are found are called *Fossiliferous.*

17. The *Direction of the Strata* or layers would be horizontal and parallel to each other, but for the disturbing forces of the earth's interior, which have raised the strata in parts, giving them uneven or inclined positions.

18. Where the strata are horizontal, which of them was the most recently formed? Which was first formed? What can you say of the heat of the earth's surface? Of the earth's interior? What can you say of the material which forms the earth's surface? What is the difference between aqueous and igneous rocks?

19. *Each Stratum of Hard Rock is Composed* of what had been soft mud, loose gravel, shells, vegetable and animal bodies.

20. The *Forms of Animal Bodies* in one stratum have been found to differ from those in the stratum below or above it, proving that at successive periods there lived successive species of animals.

The Form and Surface of the Earth.

Section IV.

THE FORM OF THE EARTH.

1. The *Form of the Earth* is that of a "*Globe*," or "*Sphere*." For this reason the topmast of a ship approaching us is first seen, then the sails, and, lastly, the body of the ship.

2. If you look around when at sea, or on a plain, what kind of a line limits your view? What is the name of that circle?

3. If you sail or move from one place to another, does your horizon change? If you go to the top of a mountain, or any eminence, how is the extent of your horizon affected?

4. *Who can see an Approaching Ship first*, the man at the foot, or the one at the top of a mountain? Which has the more extended horizon?

5. Which of these two men can first see the sun rise in the morning? Sun set? Is the day longer to one than to the other? To whom? Why does the light on a distant light-house appear to be on the surface of the water?

6. The *Continents, Islands, and Mountains* which we now behold were not formed at once; some parts were raised suddenly, but most of the land elevations were the work of ages.

7. The *Inequalities of the Earth's Surface* are no greater, relatively, than the roughness on the surface of an orange; and, although appearing to the careless observer as accidental and meaningless, they exert, nevertheless, important influences upon the conditions of mankind, and are in accordance with the wise designs of the Creator.

8. *One-fourth of the Earth's Surface* is land; three-fourths, water. In other words, the internal forces have thus far caused the elevation of one-fourth of the ocean's bed.

9. As the *Bed of the Ocean along the Coasts* is inclined, what would be the effect of an increase in the volume of water upon the size of continents and islands? Upon their elevations? What would be the effect upon the same if the volume of water should be diminished? What, if the ocean's bed should be suddenly depressed? Elevated?

sions of the earth's crust; and, in its unevenness, it is like the land above the water level.

11. The *Ocean acts an Essential Part* in the unfolding of the Creator's design to benefit mankind. It is not only the highway between the nations of the earth, but also the modifier of climate, and the vast reservoir whence the land receives its entire supply of water for the support of all life, whether animal or vegetable.

12. *If the Ocean covered the whole Surface* of the earth, could man exist?

13. *If the Surface consisted entirely of Land*, the absence of water would forbid the existence of mankind; for all vapor, clouds, rain, springs, streams, and lakes would disappear. All the fresh water of the land is raised from the great reservoir, the ocean, by the combined agencies of the sun and air, acting like a great pump and sprinkler.

14. *At the Earth's Surface* there are in contact three elements,—water, land, and air; to deprive man of any one of these would be to deprive him of life.

15. The *Earth covered with Land and Water*, but without the atmosphere, could not be the abode of man, for there would be no water to drink, no air to breathe; the land, not watered by dews and rain, could not yield him food.

16. Therefore, *Two Indispensable Agents* are provided,—the sun and atmosphere.

The *Sun by his Powerful Light and Heat* so acts upon the sea that thin, fresh water called vapor is separated from it. The vapor, like a feather loosened from a bird, is borne upward by the atmosphere, and carried far away by the winds.

Vapor becomes Clouds, and afterward returns to the earth in the form of rain, dew, or snow, to water and fertilize the soil, and to scatter all over the land innumerable springs, streams, and lakes of delicious water.

17. It is evident, then, that *All Parts of the Earth*, above and below its surface, are made to harmonize and coöperate with each other as an organized whole, for the great object of the gradual perfection of the human race.

If there were no ocean, would there be any rivers or springs? Any rain or

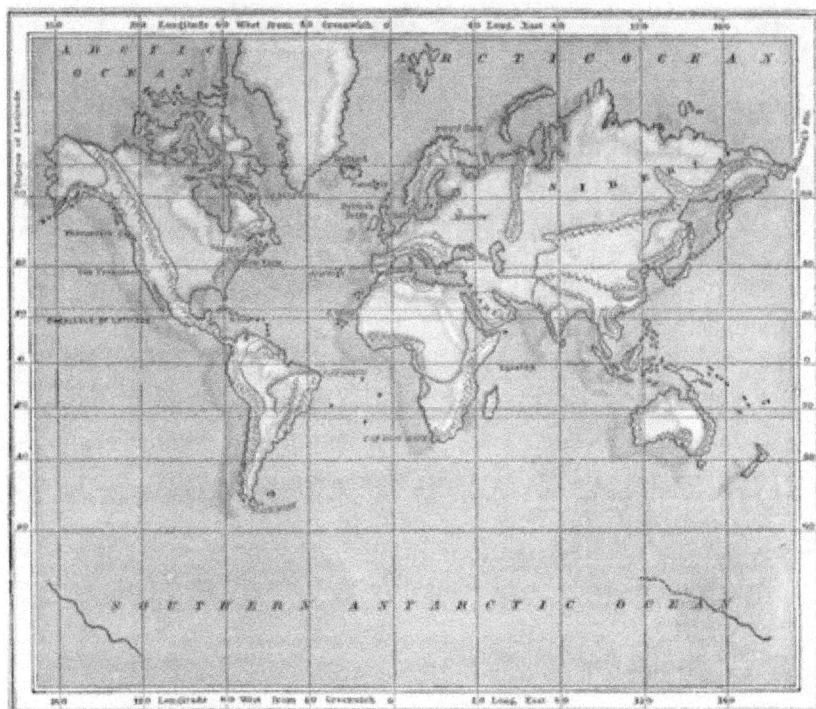

Mountain Systems are Colored Brown; Plains and Valleys, Green.

Section V.

THE CONTINENTS,—THEIR FORM.

1. The *Land on the Earth's Surface* is known, generally, as continents and islands: the continents are two in number; the Eastern or Oriental, called the Old World, and the Western or Occidental, called the New World; the islands are numerous. Australia is sometimes called a continent.

2. *When Land first emerged* from the water and came into contact with the atmosphere, it was not then as it is now, either in extent or form.

3. *None of those Large Bodies of Land* appeared, whose shapes we now trace on the globe or map; but, comparatively small points were projected, which gradually rose higher and extended more widely, according to the pressure of the forces

4. *A Continent* is entirely surrounded by water.

5. *A Continent*, with its peninsulas, highlands, lowlands, lakes, and rivers, is like a great tree that has grown from a small shrub.

6. *What is now a Vast Continent* was, at a remote period, entirely below the level of the sea; its general shape was the same then as it is now.

7. *A Continent was not raised at once*, but slowly; appearing above the water in parts.

8. *These Parts, after remaining at the Surface* for many centuries, were again submerged, and their great masses of vegetation,—trees, shrubs, and plants,—became covered over with gravel and sand.

9. *At the End of another Long Period*, the submerged vegetation and the over-lying beds would be again raised, only to undergo a similar process.

man; and although to the uninformed they appear without purpose or use, they have, nevertheless, successfully contributed toward the unfolding of God's wise design in his preparation of the earth for the abode of the human race. These vegetable masses are now the exhaustless beds of coal which supply indispensable aid to the industry and comfort of man.

11. The *Wisdom of this Plan* is further recognised in the fact that coal is found, mainly, in those parts of the earth that are best fitted for human habitation;—in the United States, Great Britain, Western Europe, British America, and China.

The Parts of the Map shown in White represent the First Land of the United States. The Parts in Dark Shading along the Coasts remained under Water until a more recent Period. The Dark Shading Inland were vast Tracts of Marsh and Woodland, but now they are the great Coal Fields of this Country.

12. The *Extended Lines of Elevation* which we call mountain chains or ranges seem to constitute the frame of the continents.

13. The *Slopes, Plains, and Valleys* have been shaped and fertilized by slides of great ice formations of former ages, and by frequent rains, which have washed down the dissolved and pulverized rocks, and the long decayed vegetable and animal substances; mixing them all together in a rich compound called mold, which supports the vegetation of the earth.

14. The *Great Body of Land Surface* is north of the Equator, both in the Old and in the New World, and comprises the whole of Asia, Europe, North America, Northern and Central Africa, and the northern part of S. America. South of the Equator are only three considerable tracts of land ; the central and southern parts of South America, the southern part of Africa, and the island of Australia.

15. The *Land of the Two Continents* not only lies chiefly in the Northern Hemisphere, but it also widens toward the north, and narrows into peninsulas at the south, these peninsulas, also, terminating in capes pointing southward, thus giving each continent the appearance of a triangle with the apex toward the south.

16. *This Peculiar Feature makes it appear* as if the water of the ocean had originally issued in great currents from the region of the Southern Ocean, as a center, and washed away

the land until arrested by the mountains and highlands of the Northern Hemisphere.

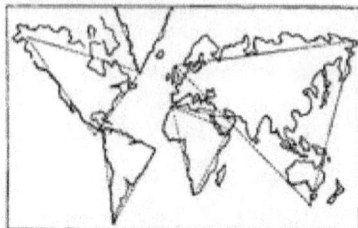

The General Form of each of the Land Divisions is that of a Triangle, the Apex pointing toward the South.

17. *Upon the Western Continent* the water seems to have encroached from the south and south-west to the foot of the vast mountain ranges which line its coast; upon Africa to the Kong and Snow Mts., and the highlands intervening ; upon Asia to the Himalaya and the Ghauts Mts.

18. *With New Zealand as a Center*, describe a great circle upon the globe, dividing it into hemispheres ; one will contain nearly all the land on the earth's surface, while the other will be composed almost entirely of water. These are known as the Land and Water Hemispheres. At or near the center of the Land Hemisphere are the British Isles. (See Map, p. 12.)

19. *By means of the Winds and Waves* new coasts have been formed, and others washed down to the ocean's bed ; loose sand on some sea-shores is carried inland, forming drift-sand hills, such as those on the southern shore of Long Island and the eastern shore of New Jersey. In some places, these movements of the sand have been attended with destructive effects, by covering houses, farms, and villages.

Tower of a Ruined Church on the East Coast of England.

20. *An Increase of the Volume of Water* would be followed by an overflowing of the land, beginning with the lowlands : thus effecting entire changes in the sizes and forms of continents.

21. The *Eastern Continent* comprises Europe, Asia, and Africa; the *Western*, North and South America.

22. The *Eastern Continent extends* in an easterly and westerly direction. Its great mountain system, commencing at Behring Strait and the Pacific Ocean, runs through central and southern Asia, and along the north and south sides of the Mediterranean Sea to Portugal in Europe, and to Morocco in Africa.

23. *These Mountains are included,* chiefly, between the parallels of 25° and 50° north latitude.

24. The *Western Continent takes its direction* from its great mountain system, which reaches from the Northern to the Southern Ocean in a north-westerly and south-easterly direction. Each of these two mountain systems is like the back-bone, which gives position and strength to an animal body.

25. The *Principal Sections* of the mountain system on the Eastern Continent are the Himalay'a, Altai (*ahl-ti'*), and Stanaroy ranges of Asia; the Cau'casus, Carpathian, Alps, and Pyrenees of Europe, and the Atlas Mountains of Africa.

26. The *Sections of the great Mountain System* of the Western Continent are the Andes of South America, and the Rocky, Sierra Madre (*se-er'rah mah'dray*), Sierra Nevada *(nay-vah'dah),* and Cascade of Northern America. These great ranges form the western defenses of America against the advance of the Pacific.

27. *On the Eastern Side of North America* is the Appalachian System, reaching from the Southern States to the Gulf of St. Lawrence, and giving to the east coast of North America its principal direction north-east and south-west.

28. *On the Eastern Coast of South America* the mountains of Brazil run parallel with the Appalachian System of North America, and secure a parallelism between their corresponding coast lines; namely, that from Newfoundland to Florida Strait, and that between Cape St. Roque and the Strait of Magellan.

29. The *Western Continent is laid out* in two great triangles, North and South America. *(See Illustration on page 11.)* Greenland has a similar shape. This peculiarity is also noticeable in the Eastern Continent, concerning its peninsulas; Africa, Hindoostan', Farther India, Corea, Kamtschatka *(kam-chat'kah),* Italy, and the Scandinavian peninsula, comprising Norway and Sweden.

Roque? Of the coast of Africa from Cape Verd to Cape Good Hope? Of the western coasts of Hindoostan' and Farther India?

Mention the principal coast lines which are parallel with each other, and have a north-easterly and south-westerly direction; those which have a north-westerly and south-easterly direction.

31. Hence, it is observed that the *General Directions of Coast Lines* are but two; namely, from north-west to south-east, and from north-east to south-west.

32. Refer to the *Maps* and you will see, furthermore, that such are the directions of nearly all the coast lines of the large islands, peninsulas, and groups of islands in the most important seas, gulfs, bays, lakes, and rivers.

33. *Australia is enclosed* by a coast line composed of six sides, all of which point in one or the other of these two directions.

34. *Above the Parallel of 40° N. Latitude* are the greater parts of North America and Asia, and nearly all Europe; while below the parallel of 40° S. Latitude extends no part of the Eastern Continent, and only the southern extremity of the Western Continent.

35. *Toward the North Pole the Land extends* and expands, as if the Southern Hemisphere was to be surrendered to the ocean; and as new land is being constantly formed in northern latitudes by volcanic action, in time the Northern Ocean may become a land-locked sea.

36. The *Arctic Ocean is connected with the Pacific* by Behring Strait, less than sixty miles in width. Indeed, the Aleutian Isles, which even now reach from Alaska to Kamtschatka, may soon, by means of their fifty active volcanoes, become a continuous rock, joining the two continents, and thus cutting off communication between the Pacific and Arctic Oceans.

37. The *Space between Greenland and Norway,* or between Greenland and Scotland, is no greater than that over which the Aleutian Isles are now being extended. It has already its stepping-stones of Iceland, the Faroe, Shetland, Orkney, and other isles, all of which have been raised by submarine forces yet in operation.

38. The *Longest Straight Line* that can be drawn on the land-surface of the earth would extend north-eastward from Cape Verd to Behring Strait, a distance of about 11,000 miles.

30. What is the general direction of the eastern coast of the Eastern Continent? See Map on page 10.) Of the coast from the south-eastern part of Arabia to the southern cape of Africa? Of the western coast from North Cape to Cape Verd? Of the eastern coast of Hindoostan? From the eastern shore of Greenland to the Gulf of Mexico? From Cape St. Roque to Cape Horn?

What is the general direction of the Pacific coast of the New World from Behring (*Bering*) Strait to Cape Horn? Of the South American coast from the Caribbean Sea to Cape St.

Land Hemisphere.   Water Hemisphere.

39. What division of the earth is in the center of the Land Hemisphere? What two divisions are nearly in that hemisphere? What division is almost entirely in it? What part of Asia is in the Water Hemisphere? What division extends furthest into the Water Hemisphere? In which hemisphere is the greater part of South America?

What islands in the center of the Water Hemisphere? Name the largest bodies of land in that hemisphere. In which of these hemispheres is the greater part of the Pacific Ocean,—the Atlantic,—the Indian?

Chart showing the Correspondence between the West Coast Line of the Old World and the East Coast Line of the New World.

Imagine the Old World to be moved westward till the mainland would meet that of the New World: what African gulf would be entered by the eastern part of South America? What American sea by the western part of Africa? Where would be the points of contact? Into what would the Amazon River flow? With what American peninsula would the British Isles be merged? Great Britain would be in what direction from Newfoundland?

40. *An Important Point of Difference* between the divisions of the continents consists in the comparative length of coast lines. In proportion to the extent of surface, the longest line of coast belongs to Europe, the next to North America, and the least to Africa. Europe, with but three sides bounded by water, has, proportionately, four times as much coast line as the whole of Africa; North America has three times as much as Africa.

41. *About One-third of the Entire Land of Europe* consists of peninsulas and islands; and, through the medium of numerous arms of the sea, this division *receives* and *bestows* strength, power, and prosperity; while the closed doors of the African coast forbid entrance to vast regions yet unexplored.

42. *To its remarkably Irregular Coast Line*, together with its mild climate and position on the globe, does Europe owe its greatness among the divisions of the earth.

43. Except in the north, *Africa has no such important Inlets* from the ocean, as those of Europe, North America, and Asia.

44. *Seas, Gulfs, Bays, and Lakes are most numerous* within a belt around the earth, embraced between the parallels of 30° and 60° north latitude.

45. *This Belt*, which is midway between the Equator and the North Pole, *comprises* the most enlightened, powerful, and progressive nations of both continents; here the human race had its origin, here is the birth-place of Christianity, and here flourished nations renowned in ancient history, which were

Therefore, the superiority of the land divisions of this section is owing, mainly, to the *influence* of their form, position within the North Temperate Zone, and the distribution of their inlets.

46. *Within this Belt*, the inlets on the coasts of the United States, British America, Western and Southwestern Europe, are numerous and important.

Mention the principal bays, gulfs, and sounds on the Atlantic coast of the United States.

Mention the principal seas in Western and Southwestern Europe. Mention the principal bays, gulfs and channels.

47. The *Condition of a Race or People* is affected by contact with surrounding nations and influences; and the greater the facilities for communication and inter-communication, the greater is the advancement; hence, inlets, rivers, canals, and railroads promote the civilization and progress of man.

A City.—Ruin.—Harbor.—Railroad.—Commerce.—Agriculture.

48. *Asia and Europe together form* a vast peninsula, which, with that of Africa, composes the Eastern Continent.

49. *Were it not for a Separation of Sixty Miles* between the Mediterranean and Red Seas, each of these peninsulas would be a vast island or continent.

In this respect, what similarity exists between the Old and the New World? Were the Isthmuses of Darien and Suez overflowed, how many and what continents would there be?

50. The *Peninsula comprising Europe and Asia* has its greatest extent from Behring Strait on the north-east to Portugal in the south-west, a distance of about 8,500 miles, or one-third the earth's circumference. It is remarkable for the number and extent of its indentations, which give to it the appearance of a great plant, extending its numerous roots in all directions for nourishment and strength.

51. *This is not the case*, however, with South America, and still less with Africa, which is like a plant almost destitute of roots.

Mention the principal indentations of Europe; the peninsulas formed by

52. *Europe extends* from the foot of the Ural Mountains westward, over a great expanse of land,—a continuation of the northern plain of Asia,—to the Carpathian Mountains and the Baltic Sea. Beyond these limits it becomes narrow; facilitating external and internal communication.

53. The *Coast Line is so greatly diversified* by the penetrating arms of the Atlantic Ocean and the Mediterranean Sea that nearly all western and south-western Europe is composed of peninsulas.

TABLE SHOWING THE COMPARATIVE EXTENT OF COAST-LINE.

GRAND DIVISIONS.	SQUARE MILES.	LENGTH OF COAST LINE.	SQUARE MILES FOR 1 OF COAST
Europe	3,800,857	17,000	222
North and Central America	9,059,927	24,000	345
South America	6,954,131	13,800	477
Asia	16,415,748	23,000	559
Africa	11,558,660	16,000	741

54. The *Three great Land Divisions of the South*,—Africa, South America, and Australia,—resemble each other in their lack of sea arms, and in their backwardness of development; presenting, in these respects, a strong contrast to the divisions of the North.

55. The *Western Continent has its greatest Extent* from the northern part of Russian America, south-eastward to the Strait of Magellan, a distance of about 10,000 miles.

56. The *Northern and North-eastern Parts of N. America* are remarkable for their great number of inlets from the sea, cutting the land into a great variety of islands and peninsulas.

57. *Baffin Bay separates* Greenland from the main land of the Western Continent, and *Hudson Bay* forms the great peninsula of Labrador and East Main.

58. *As you go South, you meet* the Gulf of St. Lawrence, Gulf of Mexico, Caribbean Sea, and the Mouth of the Amazon.

59. *Characteristic of the Atlantic Coast of the United States*, are its numerous bays and other inlets; the principal being the Chesapeake, Delaware, New York, Narragansett, Massachusetts, and Penobscot Bays; besides Long Island, Pamlico, and Albemarle Sounds. On *the Pacific Coast*, the most important inlets are San Francisco Bay and Puget's Sound.

60. *South America has its entire North-eastern Side turned* toward Europe and North America, as if to invite their aid in its development; and, although joined by land to North America, the water affords far easier communication than the mountainous region of the isthmus.

61. *Had the Wide Pacific rolled between Europe and America*, instead of the narrow Atlantic, Columbus would probably not have discovered America; or, *had the great Mountain System* of America been placed on the eastern coast, shutting out the Atlantic as it now does the Pacific, and presenting to the east the same undeviating coast line that it does to the west, the New World would probably be less adapted to the progress of mankind than Africa or Australia.

62. *Between the eastern side of the New World and the western side of the Old*, there is a remarkable analogy, not only in the parallelism of the general coast lines, but also in their system of seas, bays, and other inlets from the ocean.

Section VI.
THE CONTINENTS,—THEIR RELIEFS.

1. The *Land of the Continents* is greatly diversified,—low in some parts and high in others; the altitude or absolute elevation of a place being the distance above the level of the ocean.

2. The highest mountains, as compared with the size of the earth, are no larger than grains of sand on a globe ten inches in diameter; they nevertheless exert vast influences upon the conditions of the whole land surface of the earth.

3. *Plains elevated but slightly* above the level of the sea are called lowlands, even though hills may rest upon them; those of higher elevations, enclosing and supporting mountains, are highlands or plateaus.

4. The *Transition from Low to High Land* is varied; being either abrupt, gradual, or terraced.

5. *A Mountain Range or Chain* is a succession of mountains which have similar geological formations. The *Highest Point* in a chain is called the culminating point.

6. *A Mountain System* is two or more parallel ranges, connected with each other, or which rest upon the same plateau.

7. The *Soil of the Valleys* is fertile, and the climate generally delightful.

A Valley in Switzerland.

8. Although *Mountains and Plateaus* are both elevations of land, and are connected, yet they should be considered distinct from each other. The rugged, broken outline of lofty mountain peaks, with their intervening valleys and passes, presents a strong contrast to the comparatively dull and even surface of a plateau; just as a deeply indented coast does to one whose line is almost unbroken.

9. *No Precise Height* has ever been prescribed, according to which elevations of land should or should not be called mountains.

10. The *Loftiest Peaks on the Globe* are among the Himalayas, the principal one, Mt. Everest, being over 29,000 feet high. Mt. Aconcagua, the highest in S. America, is 23,906

The highest peaks of the Rocky Mts. are between 13,000 and 15,000 feet high. The White Mts. are about 6,000, the Catskills 3,000, and the Alleghanies from 1,000 to 5,000 feet.

11. *A Plateau is* an extent of land elevated above the level of the sea from 2,000 to 14,000 feet.

12. The *Surface* may be level, rolling, or hilly ; some plateaus contain mountains, valleys, and lakes.

13. *Plateaus owe their Elevation* to internal forces, exerted, not as in the more sudden and violent formation of mountains, but slowly and gradually ; giving them a comparatively level and unbroken surface. Should, however, the force from beneath be so violent as to cause *Openings or Seams* in the earth's crust (see Illustration, page 8), there would be projected through this fissure melted mineral matter, called lava, besides stones, cinders, and ashes ; which, falling and hardening upon the uplifted surface, would form a conical pile called a mountain.

14. The *Upheaval of Hills and Mountains from the Bottom of the Sea* accounts for the finding of sea-shells on their sides and tops ; and the boulders, stones, pebbles, and gravel found in all countries, were irregular fragments of rock, broken off by violence or by atmospheric action, and carried great distances by the rush of water, ice, and icebergs, from high to low ground.

15. *Mountains were raised to their Present Elevation* by violent and repeated convulsions, the process extending over thousands of centuries. It is the opinion of geologists that the upheaval of the highest mountains was more sudden, and attended with more violence than that of the ranges of less elevation ; that the Alleghany and Brazilian Mountains were raised more slowly, and in earlier periods, than the Rocky and the Andes Mts. The Alps were upheaved more suddenly, and at a period comparatively recent.

16. *Mountains which have been violently Elevated are known* by their deep fissures, and great displacement of strata and fossils.

17. The *Direction of a Chain of Mountains* is due to the position of the rent made in the earth's crust.

18. *Mountain Chains extend* mostly in either of two general directions ; from north-east to south-west, or from north-west to south-east.

19. What chains extend from north-east to south-west ? What from north-west to south-east ?

20. The *Pressure from beneath forces up,* also, masses of the earth's crust from a considerable depth. Granite is sup-

23. The *Rocky,* the *Andes,* and the *Scandinavian Mountains* have their long and gradual slope on the east, and descend abruptly on the west. The *Himalayas* and the *Alps* descend abruptly toward the south. The highland surface of Spain is terraced from the Pyrenees and Cantabrian on the north to the Strait of Gibraltar on the south.

24. The *Great Plateau System* of Asia lies south of the Altai Mountains ; that of Europe south of the Baltic Sea ; of Africa south of its central part ; and of America along the west coast.

25. The *Climate on Mountains and Plateaus* is cooler than on the lowlands of the same latitude, and the greater the elevation the lower the temperature : hence, upon the *Elevation of a Country,* as well as upon its latitude, depend its climate, productions, and to some extent, the pursuits of the inhabitants.

26. *Elevated Regions serve to moderate* the temperature of the lowlands adjoining them. When air is heated it becomes lighter than the cooler air above it, and ascends ; the cold air descending to take its place.

27. Therefore, as the *Elevations are greatest in the Hot Regions* of the earth, and diminish toward the poles, the inhabitants of the sultry tropical plains, at the foot of lofty mountains, are continually refreshed by the cool air which comes down from their snowy summits.

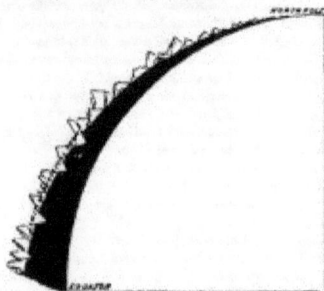

Comparative Height of the Mountains in America, from the Equator to the North Pole ; also, the Limit of Perpetual Snow

28. For the same reason that you put a piece of *Ice into a Pitcher of Water* in summer, rather than in winter, Providence has uplifted the highest mountains in the tropical, and not in

32. *Nearly the whole of Mexico* is a plateau, whose inhabitants, even in the tropical part of the country, enjoy a temperate and healthful climate, owing to its great elevation above the sea.

Section of Mexico from the Pacific Ocean to the Gulf of Mexico

33. The *City of Mexico* is 7,400 feet above the sea level, about twenty times higher than Trinity Church steeple, in the city of New York.

34. *Central and Southern Africa* is one vast table land, the most extensive in the world. It descends on all sides by terraces, to the strip of low ground along the coast.

35. The *Great Mountain System of Europe*, comprising the Cau'casus, Alps, Pyrenees, Cantabrian, and Apennines, is in the southern, or warmest part of that division.

36. *In the Northern Regions of Europe* the only important elevations are the Scandinavian Mountains of Norway and Sweden, which, however, average less than one half the height of the mountains in the south of Europe. With this exception, the northern regions are, comparatively, lowlands.

37. *Take away these lofty Mountain* ranges and extensive plateaus from the places now occupied by them, or remove them from the hot to a cold zone, thus increasing the heat of the tropical and the cold of the frigid regions, and the consequence would be a complete derangement of climates, productions, and the conditions of the inhabitants.

The Alps.—A Glacier.—A Tunnel in the Ice whence issues a Stream which is the commencement of a Large River.

38. *In the Tropical Andes, the Region of Perpetual Snow* is above the line of 16,000 feet elevation ; *in the Alps,*—Tem-

perate Zone,—it is about 8,500 feet above the sea level ; and, *in Arctic Latitudes,* it reaches down to the sea.

39. The *Masses of Snow upon the Mountains* being constantly increased, force their way down the valleys to warmer regions below the snow-line. *By Pressure, alternate Thawing and Freezing* of the upper surface, the whole becomes a great stream of ice, called a *Glacier,* varying in depth from a few hundred to several thousand feet. The *Water that descends through the Crevices* of the ice unites with springs and flows down the mountain sides through tunnels which it cuts in the ice and snow. Every glacier is thus the source of a stream. The *Best-known Glacier Region* is that of the Alps.

40. *Draw a Line from the Sea of Mar'mora Northeastward to Behring Strait,* and you will have, south of this line, nearly all the great elevations of Asia, consisting of a vast system of plateaus, supporting lofty mountains whose tops are constantly covered with snow ; to the north of these lies the great Siberian Plain.

41. The *Highest Plateau on the Globe* is that of Central Asia, which extends 1,500 miles from the Altai Mountains on the north, to the Himalayas on the south, and 2,500 miles from west to east ; having about the same dimensions as the United States, and an average elevation above the sea of 10,000 feet. Its *Surface is greatly diversified* with heights and depressions, rivers and lakes. The principal rivers are the Ganges, Brahmaputra, Indus, Amoor, and Hoang Ho. The Ganges has its two principal sources situated in immense masses of snow, at the elevation of 13,000 feet. The *Elevations diminish* gradually from the Himalayas northward to Siberia, where the slope continues downward to the Arctic Ocean.

42. *Nearly all Western and South-western Asia* consists of plateaus about 4,000 feet high.

43. *This System of Highlands extends* westward to the Atlantic Ocean, over Southern Europe and Northern Africa ; the Mediterranean, Caspian, and Black Seas being considered its great depressions.

44. While much *the larger part of Asia consists* of vast plateaus, *Europe consists* mainly of an extended plain, which commences at the Strait of Dover, extends eastward between its great mountain system and the Baltic Sea, and then opens upon and covers Russia. The surface of this plain is almost level, and has an elevation of about 1,000 feet.

45. The *Average Height of the Alps* is between 8,000 and 10,000 feet ; the highest peak, *Mt. Blanc,* being over 15,000 feet. The *Apennines* average from 4,000 to 8,000 feet ; the *Sierra Nevada* of Spain from 6,000 to 10,000 feet ; and the *Scandinavian* Mountains of Norway and Sweden about 4,000 feet.

46. The *Great Plateau of Africa* ranges from 2,000 to 10,000 feet in elevation ; its highest part being in Abyssinia.

47. The *Loftiest Peaks in Africa* are Kenia and Kilimandjaro, whose summits are 20,000 feet above the sea.

48. *Central Africa,* north of the Equator, descends to the level of the Great Desert, which is between 1,000 and 2,000 feet above the sea. The highest ranges on the African plateau are the Abyssinian, Cameroon, and Snow Mountains ; the highest peaks are Kenia and Kilmandjaro.

49. The *Principal Plateaus of the New World* are in South America, among the Andes.

Comparative Elevation of Cities, Mountains, and Lakes.

The Andes.   Llanos and Pampas.   Brazilian Mts.
The Rise of South America.

58. *As the Torrid Regions of the Earth require the greatest amount of Rain*, there are the loftiest mountains, which act as huge condensers of the clouds and vapors floating in the atmosphere; and by the melting of the snow on their sides, they supply springs and rivers to the plains below.

59. *If South America contained no such Elevations*, the quantity of rain poured upon the vast plains would be greatly diminished.

60. In the tropical regions of South America the *Rain-bearing Winds blow*, not from the Pacific, but from the Atlantic Ocean. The clouds, floating westwardly over the land, feel the cooling influence of the Andes, and respond with copious rains, which cover with the heaviest vegetation a region that would otherwise be a sunburnt wilderness.

61. *In Some Districts between the Andes and the Pacific*, rain is almost or wholly unknown, because the clouds are exhausted before passing the mountains.

62. *Had the Andes been raised on the Eastern Side* of that great peninsula, instead of on the western, the rain would fall in torrents upon the then short Atlantic slope, and South America would be deprived of its immense rivers, dense forests, and fertile plains.

63. *Although the Mountain Chains and Plateaus of South America are Extensive*, yet they only cover about one-fifth of its surface, the greater part of it being vast plains.

50. The city of *Quito*, (ke'to,) in Ecuador, is built on a plateau nearly 10,000 feet above the Pacific.

51. *Potosi*, a city of Bolivia, is built on a plateau so high that the streets of the city have an elevation of more than 12,000 feet above the level of the sea.

52. *Lake Titicaca*, (tit-e-kah'kah,) between Bolivia and Peru, has nearly the same level, being twice the height of Mt. Washington in New Hampshire, four times that of the Catskill Mountains, and seven times that of the Blue Ridge at Harper's Ferry.

53. *High as are these Cities, Lakes, and Table-lands*, yet they are far over-topped by the surrounding mountains, which rise about 10,000 feet above them; hence, these places are but little more than half-way up the highest of the Andes.

54. The *Rocky Mountains, if placed beside the Andes*, would reach only to the plateaus of the latter. The elevation of the *Appalachian* range is only about one-seventh that of the Andes.

55. *On the Western Side of the Andes*, the slope toward the Pacific is abrupt: on the eastern, or Atlantic side, it is gradual; being interfered with only by the Brazilian Mountains, which, however, are less than one-fourth the height of the Andes.

56. The *Andes rise so High* that their tops are in the region of perpetual snow, while, at their foot, the heat is oppressive, and would be greatly intensified, but for their cooling

64. *These huge Piles, called Mountains, projected by Violence* through fearful gaps in the earth's crust, from the melted interior, and occupying such positions of usefulness to the earth and to man, stand in their appointed places, as monuments, not of the Creator's power alone, but also of His wisdom and goodness.

65. *By means of these great Upheavals*, man derives a knowledge of the interior formations of the earth, and obtains

67. *Among the many remarkable Features* in the formation of mountain ranges, is one that deserves notice on account of its bearing upon civilization; it is *their Formation in Peaks*, between whose sloping, or perpendicular sides, *Passes* are left. A chain of peaks, resting on a plateau, is termed by the Spaniards, *sierra*, from its resemblance to a *saw*.

68. *Hold up your Hand* and you will have a good illustration of a section of mountains and a plateau; the fingers, separated from each other, represent the mountain peaks, and the hand represents a plateau.

69. *If the Continents were deprived of their Land Elevations*, the change effected in the climates alone would render the now fruitful plains unfit for the abode of mankind.

70. *Were the Tops of high Mountain Ranges connected*, so as to form a continuous barrier, nations on opposite sides of the chain might be further apart in their relations with each other than if an ocean rolled between them.

71. *In most of the Great Chains there are Natural Passes* far below the summits of the mountains.

72. The *Passes through the Alps* are not half-way up the mountains; they are proportionately lower than those of other leading chains.

73. The *Advantages of National Communication* are now seen by man, but they were recognized by the Creator when he formed the mountains, with their intervening passes.

74. *If the whole Land Surface of the Earth were made Level* by filling up the lowlands with the material from the elevations, its height would be about 900 feet above the sea level.

75. *If the Matter comprising all the Mountain Systems of the World* were transferred to the polar regions, they would not be sufficient to make the polar diameter equal to the equatorial.

76. *In North America there are Two great Mountain Systems;* the Rocky and the Appalachian, or Alleghany.

77. The *Rocky Mountain System is supported* by the North American Plateau, which is elevated from 4,000 to 7,000 feet, and extends over a great part of Central America and Mexico, the western third of the United States, and the western part of British America.

78. *In this System are included* the Cascade Range, Sierra Nevada, and Sierra Madre. East of the Sierra Nevada is the Great Basin, or Plateau of Utah.

79. The *Rocky Mountain System extends* from the Isthmus of Panama, in a north-westerly direction, to the Arctic Ocean, at about 70° north latitude.

80. *Its Widest Part is* in the United States, and embraces all that region between the Pacific Ocean and the central part of Colorado, a distance of over 900 miles.

81. The *Surface of the Plateau* slopes eastward from Pike's Peak to the Missouri River.

82. The *Most Western Range* of this system commences at the southern extremity of Lower California, and extends along the Pacific coast as far north as Mt. St. Elias, in latitude 60°.

83. The *Sierra Madre commences* at the southern part of Colorado and extends into Mexico.

84. The *Most Northern Pass* in the United States through the Rocky Mountains, is near the head waters of the Missouri and Lewis Rivers, and is one of the routes proposed for a railroad to the Pacific.

85. TABLE SHOWING THE CULMINATING POINTS, AND THE MEAN ELEVATION OF THE LAND.

	Mean Elevation. Feet.	Culminating Points.	Feet.
Asia	1,000	Mt. Everest	29,000
South America	1,060	Mt. Aconcagua	23,205
North America	700	Vol. Popocatepetl	18,500
Europe	680	Mt. Elborus	17,800

The Mountains and Plateaus of the United States are here colored Brown; the Lowlands and Valleys, Green.

Section VII.

VOLCANOES,—EARTHQUAKES.

A Volcano, and Fissures caused by Earthquakes, may be illustrated by means of a Cake which is burst open at the Top, by the Escape of Steam arising from the Fluids within the Cake; the Heat of the Oven corresponding to that of the Earth's Interior.

1. *Volcanoes, Earthquakes, the Rising and Sinking of the Land* are all attributed to the pressure of steam and gases, proceeding from the heated interior of the earth.

2. *A Volcano* is an opening in the earth's crust through which issue melted rock, or lava, stones, ashes, flame, smoke, and steam. (*See Illustration on page 8.*)

3. The *Materials thrown out usually accumulate* around the opening, called the crater.

4. A *Rent in the Earth's Crust* may be made beneath the sea, where a high mountain will sometimes be formed; sometimes no elevation appears; the fire, lava, and other material being thrown upwards through the water.

5. *Volcanoes allow the Escape* of fire and gases from the interior of the earth, and thus prevent greater destruction by earthquakes.

6. *Some Volcanoes remain inactive* for long or short periods; some now called extinct may again become active.

7. Volcanic action is usually preceded by earthquakes, which sometimes rend the earth open in fissures, and engulf whole villages and cities.

8. *By these Convulsions Mountains and Hills are raised,* in some instances, from what, a few hours before, were low

one of the Lipari (lĭp'a-re) Islands, Hecla in Iceland, Cotopaxi (ko-to-pax'e) one of the Andes, Sangay near the city of Quito, Mauna Loa on the island of Hawaii (hah-wi'e), and Teneriffe on one of the Canary Islands.

Fissures caused by an Earthquake in Italy 1703.

12. The *Number of Active Volcanoes on the Earth* is about 250, more than half of which are on the coasts and islands that line the Pacific Ocean. The most remarkable volcanic region is in Malaysia. Continents have their volcanoes mostly on their borders; those of the Western Continent are chiefly among the Andes and the Rocky Mountains.

13. Although *Earthquakes mostly occur in Volcanic Districts*, yet any part of the earth's surface is subject to them. Some are violent and destructive, while others are almost or entirely imperceptible.

14. *On the Western Continent, Earthquakes are most frequent* in Central America, Chili, and Peru; in Europe, they occur chiefly in Italy and its vicinity.

15. The *Approach of an Earthquake*, like the eruption of a volcano, is sometimes indicated by symptoms of unusual agitation beneath the surface of the ground.

16. *Among the Greatest Earthquakes* of which we have a record, is that which destroyed the cities of Herculaneum and Pompeii (pom-pay'e), A.D. 63; and, after they had lain in ruins for sixteen years, they were again overwhelmed by an eruption of Mt. Vesuvius.

17. In 1692, *Port Royal, the Capital of the Island of Jamaica*, was sunk in less than one minute; the sea rolling in, and driving the vessels that were in the harbor over the tops

in the fissures which opened, and immediately closed over them. A *Portion of the Earth four times as large as Europe* was affected by this terrific shock. The *Waters of the Scotch Lakes* suddenly rose above, and then subsided below, their level. On the *Shores of the West Indies* the tide rose twenty feet, and the water resembled ink; even the coast of Massachusetts and the waters of Lake Ontario were sensibly affected.

19. In 1811, occurred the *Earthquake of New Madrid*, in Missouri, which was remarkable for the continuous quaking and rending, over an extent of 300 miles, during several months. *Great Openings* were made in the surface, from which mud and water were projected.

20. *These Internal Convulsions continued* until they culminated, March, 1812, in the *Earthquake of Caracas*, on the northern coast of South America, by which the whole of that splendid city became instantaneously a mass of ruins, and thousands of its inhabitants perished.

21. In 1822, *an Earthquake occurred in Chili*, which resulted in the elevation of a large section of country to a height varying from two to seven feet.

22. In 1857 and 1858, *Repeated Shocks were felt*, at intervals, in the country around Naples. Several towns were reduced to heaps of ruins, and about 30,000 inhabitants perished.

23. *During the Earthquake*, Mt. Vesuvius continued in action; and, by affording a means of escape for the confined gases, doubtless prevented the entire destruction of the city of Naples and the ruin of all the region in the immediate vicinity of the volcano.

24. In 1859, *the City of Quito* (ke'to) and several towns in its vicinity were almost entirely destroyed by an earthquake.

25. In August, 1868, *an Earthquake occurred in Peru, Chili, and Ecuador*, which caused a fearful loss of life and property; and, in October, several shocks were experienced in California, causing considerable damage in the principal cities.

26. *Shocks have been felt* at different times in various parts of the United States.

27. *Subsidences, like Upheavals*, sometimes occur so gently that the inhabitants are only aware of the change by the difference in the sea level. In 1819, an area of 2,000 square miles about the mouth of the Indus, in Hindoostan, was suddenly converted, by an earthquake, into an inland sea.

Fort Sindree before it was Submerged by the Earthquake of 1819.

Fort Sindree after the Earthquake.

28. The *Fort and Village of Sindree sank* so much that only the tops of the fort, houses, and trees were seen above the water.

29. The *Coast of Sweden* has been rising for many years; near Stockholm, at the rate of a few inches in a century.

30. *In Greenland*, the south-west coast has been slowly sinking for four centuries past.

31. In 1866, *an Island was Upheaved* from the bottom of the sea south-east of Greece. The water was violently agitated, and from the fissures rushed flame, smoke, lava, and fragments of rock.

Section VIII.
PLAINS AND VALLEYS.

1. The *Land Surface of the Earth* may be divided into two general classes, highlands and lowlands; the highlands, comprising mountains and plateaus; the lowlands, plains and valleys.

2. *Lowlands comprise* all lands whose elevation is not more than 1,000 to 1,500 feet above the sea.

3. *A Plain surrounded by Mountains* or hills is called a valley.

4. *Through the Lowest Part of a Valley*, or near its center, generally flows a river, which drains it.

5. *Lowlands comprise* far the greater part of the land surface of the earth, and in them is found the great mass of vegetation, animals, and mankind.

6. The *Soil of the Lowlands* is constantly enriched by the alluvial washings from the mountain sides, which have filled the fissures and depressions of the rock that originally formed the land surface of the earth; it is still further enriched by the collection upon it of decomposed vegetable and animal substances. About two-thirds of the Western Continent are covered by plains.

7. The *Great Central Plain of North America* is all that part north of the Gulf of Mexico and between the Rocky and Alleghany Mountains, an area of about three and a quarter millions of square miles; comprising four great basins, drained

by the Mississippi, St. Lawrence, and Mackenzie Rivers, and Hudson Bay.

8. The *Lowlands of South America* comprise those of the Orinoco, Amazon, and La Plata Rivers, and cover four-fifths of the surface east of the Andes.

9. The *Lowlands of the Orinoco, termed Llanos,* are less than 300 feet above the sea level, and present a surface almost as even as that of water. *During the Dry Season,* from May to November, the ground is parched and barren; presenting the appearance of a desert. *During the Wet Season,* from November to May, the clouds, driven westward by the Trade Winds, pour down their rain; when horses, cattle, serpents, and alligators suddenly appear in vast numbers.

10. The *Plains or Lowlands of the Amazon, termed Silvas,* extend from the Andes to the Atlantic, a distance of 1,800 miles, and average 600 miles in breadth. *They cover an Area* of about 2½ millions of square miles, and consist, chiefly, of dense forests into which man has scarcely penetrated.

11. The *Plains of the Amazon* are about two-thirds the size of all Europe.

12. The *Valley of the La Plata* consists mainly of vast grassy flats, called *Pampas,* where vast herds of cattle feed; *these Animals are hunted for* their hides, horns, and tallow, which constitute the chief export of that region.

13. The *Three Plains of South America* cover an area of 5,000,000 square miles, while all Europe contains but 3,500,000 square miles.

14. The *Great Northern Plain of the Old World* lies north of its chief mountain system. *It Extends* from the shores of the North Sea and English Channel, eastward, over France, Belgium, Holland, Denmark, Northern Germany, Russia in Europe, Russia in Asia, and Independent Tartary, to Behring Strait; interrupted only by the Ural Chain, which forms a natural boundary between Europe and Asia.

15. The *Portions of this Great Plain* which are drained by the tributaries of the North, Baltic, and Black Seas, are famous for their fertility.

16. *That Part of the Plain bordering on the North, Baltic, and White Seas,* evidently emerged from the ocean at a much later period than some other parts of the continent; indeed, its elevation is yet incomplete; for many parts of Holland are still below the sea level, and are protected from inundation by means of dikes constructed by the inhabitants.

17. In the *Region of the Caspian and Aral Seas,* the surface is also much depressed; some parts being below the level of the sea.

18. *Until a Period comparatively Recent,* it probably formed the bed of a great inlet, or arm of the ocean, from which it has been isolated by the upheaval of the surrounding highlands. The soil contains sand, sea-shells, and salt, and the region is consequently desolate. There being no outlet to the enclosed waters, the seas of this basin are strongly impregnated with salt.

19. *Toward the Arctic Ocean,* the plains in Europe and Asia are a boundless waste, swampy in summer, and frozen in winter.

20. The *Polar Regions of North America* may be considered a continuation of the lowlands of Northern Asia.

Section IX.

DESERTS AND OASES.

1. *Deserts are Extensive Tracts* destitute of water, and, consequently, of vegetation and animal life.

2. *Their Condition is Attributable,* chiefly, to the heat and dryness of the winds which blow over them.

3. The *Desert Region of the Old World* extends over the greater part of Northern Africa, and north-eastward over vast regions of Arabia, Turkey, Persia, Afghanistan, Beloochistan, Independent Tartary, and the Chinese Empire; this is, also, the great rainless region of the world: its area is more than twice that of the United States.

4. The *Surface of that part of Sahara* which lies north of Timbuctoo (*see Map of Africa*), thence toward the Atlantic, is a vast sandy waste covered with a coating of salt and sea-shells.

5. At times, the *Desert is Visited by the dreaded Simoon—* a hot, suffocating wind which drives the burning sand in great clouds furiously over the surface, for great distances.

6. *To avoid Suffocation,* travelers throw themselves on the ground with their faces to the earth, stopping their ears and noses with their handkerchiefs until the storm has passed; their camels lie close to the ground and bury their noses in the sand.

7. *By means of the Winds which Blow over the Desert,* some houses, villages, and towns have been completely covered with the driven sand. *There have been Discovered* remains of ancient temples so long buried that no record of them is found in history.

8. *Large Portions of the Great Desert* are diversified by hills and mountains, between which are valleys or immense tracts either of sand or naked rock.

9. *Between Fezzan and the Southern Side of the Desert,* some tribes live on the mountains, at elevations where the temperature requires them to wear warm clothing, even furs. Here, also, rain occasionally falls; while in other districts, the mercury in the thermometer rises to 132° in the shade and 156° in the sun.

10. *Sahara is a vast Plateau* which has an elevation above the sea of 1,200 to 1,500 feet. It is about 1,000 miles wide and 3,000 miles long; covering an area equal to about four-fifths that of the United States.

11. The *Oases are* fertile spots in various parts of the desert, *where are found* springs of cool and delicious water, besides grass, the palm, fern, acacia, and other trees; here travelers and their camels find shade, refreshment, and rest.

12. The *Oases are Depressions* in the table land of the desert; the water is supplied from the surrounding cliffs, and is retained by a stratum of clay in the center of the valley.

13. The *Number of Oases in Sahara* is about thirty; of which, twenty are inhabited.

14. The *Principal Desert in the New World* is that of Atacama, where rain has never been known to fall. It is situated in Peru and Bolivia, west of the Andes. Its dry surface of sand and rock supports not the slightest vegetation.

The Ocean.—A Storm.   Some of the Uses of Water.   The Ocean.—Fair Weather.

Section X.

THE OCEAN: ITS EXTENT AND DIVISIONS.

1. The *Existence on the Earth's Surface* of a vast body of water is essential to life; for, in the composition of both vegetable and animal bodies, the chief element is water.

2. *Water forms* more than five-sixths of the animal body, and nearly the whole of the vegetable.

3. All *Lakes, Streams, Springs, Rain, and Clouds,* besides all vegetables and animals are, consequently, dependent upon the ocean, which is the great reservoir whence all the land on the earth's surface receives its supply of water.

4. *Influenced by a certain degree of Cold,* water becomes ice; and, influenced by heat, it takes the form of steam and vapor.

5. *Water exists* not only on the earth's surface, but also in the air above the surface, and in the ground below it, where it forms subterranean lakes and streams.

6. The *Water of the Ocean is preserved Pure* by its saltness and constant motion. Fresh water is that which has been raised from the ocean by evaporation, and returned to the land by condensation.

7. The *Sea or Ocean has Five Divisions,* called the Pacific, Atlantic, Indian, Northern, and Southern Oceans.

8. *It affords an Easy Communication* between nations, for their mutual development and prosperity.

9. As there are *Two Great Bodies of Land,* the Eastern and Western Continents, so there are two principal oceans corresponding to them, in both size and shape; the Pacific to the Eastern, and the Atlantic to the Western Continent. The Indian Ocean may be considered a part of the Pacific.

10. *In America, the Mountain Ranges correspond in Size* to the oceans nearest them; the Andes and Rocky to the Pacific, the Appalachian and Brazilian chains to the Atlantic.

The highest peaks of the Andes border on the widest part of the Pacific.

11. The *Largest Ocean* is the Pacific, which contains about one-half the water on the globe, and covers one-third of the earth's surface. It extends from Behring Strait to the Antarctic Ocean; its western shore being Asiatic, and its eastern, American.

12. The *Shape of the Pacific and Indian Oceans* is the reverse of that of the continents, being narrow in the north, and wide in the south.

13   AREAS OF THE OCEANS.

	Square Miles.
Pacific	84,000,000
Atlantic	25,000,000
Indian	20,000,000
Arctic	4,000,000
Antarctic	4,500,000
Total	143,500,000

14. While the Pacific is distinguished for its size, the Atlantic is distinguished for its numerous arms which penetrate far into the land of both continents.

15. *Owing to these Arms, and the Position* of the Atlantic between the important sides of the continents, this ocean contributes far more than any other to the interests of mankind.

16. Mention the principal arms of the Atlantic on its eastern side; on its western.

Into which of the grand divisions do they mostly penetrate?

In what seas are most of these arms?

Has the Pacific such arms on both sides? On which side are its principal arms? Mention them. Mention those of the Indian Ocean.

17. The bed of the sea, like the surface of the continents, is diversified by highlands and lowlands; the submarine plateaus causing shallow water, termed shoals and banks.

18. *Near some Coasts,* the ocean is shallow, its bed being the submerged border of the continent; but, at a distance from the coast of 100 to 300 miles, the water becomes suddenly deep. (*See Illustration on following page.*)

A Sectional View from the Atlantic Coast of the United States eastward and north-eastward; showing the Bed of the Ocean, the Comparative Shallowness of the Water near the Coasts, the Depth of the Ocean, as compared with that of the high and North Seas, in Summer sailing from Europe to the United States.—Fishing Vessels off the Coast.—a Wreck at the Bottom of the Ocean.—Whales.—Seaweeds.

19. The *Depth of the Water surrounding the British Islands* and the Islands east of Asia, is only about one-fortieth of that of the ocean basin.

20. If the ocean were withdrawn from the earth, its bed would appear chiefly as extensive valleys of various depths, and the parts adjoining the continents, as plateaus, sloping suddenly downward to the valleys.

21. The *Ocean is Deepest* near the tropics; here, also, are the highest mountains.

22. The *Depth of the Ocean* varies from 1,000 to 30,000 feet. Between Ireland and Newfoundland the bed of the sea is a submarine plateau, remarkable for its comparative evenness, and the quietness of the waters that rest upon it. The depth of the water there varies from 10,000 to 15,000 feet.

23. The *Depth of the Gulf of Mexico* is about 5,000 feet in its deepest part; of the Mediterranean from 3,000 to 9,000 feet; of the North Sea, 180 feet. The mean depth of the Ocean is estimated to be between 15,000 and 20,000 feet.

24. *A Depression of the Water Level* of about 300 feet would extend the main land of Europe and Asia over their neighboring seas and islands.

25. Were the *Mass of Water diminished*, so that its greatest depth would not exceed 5,000 feet, the elevation of the continents would be so increased that the climate of the lowlands, even in the temperate and torrid zones, would cause them to become frozen wastes; the most fertile plains of Europe would then have an elevation above the depressed ocean level of over 15,000 feet, the present height of Mt. Blanc; the Mississippi valley would attain a far greater elevation than the present altitude of the highest peaks of the Rocky Mountains.

26. Therefore, it is plain that *the Climate of any Locality depends* essentially, not only upon its distance north or south of the Equator, but also upon its elevation above the level of the sea.

27. The *Saltness of the Ocean* is supposed by some to be caused by great masses of salt, forming parts of its bed, or by the salt brought into it by rivers; others hold that it was originally made salt by the Creator.

Section XI.

MOVEMENTS OF THE OCEAN.

1. The *Movements of the Oceanic Waters* are of three kinds,—waves, currents, and tides. Waves may be influenced by tides or by winds. The tide affects the whole depth of the ocean; the wind affects the water nearer the surface.

2. *Currents and Tides* are regular and constant.

3. *Tides are caused* by the influence of the moon and sun; mostly of the former.

4. The *Oceanic Currents are caused*, or modified, by the winds, the difference of temperature between the Equator and the poles, and by the revolution of the Earth on its axis.

5. *If the Earth were at Rest*, the whole surface covered evenly with water, and under no external influence, there would be no currents, or important movements of the water; but admit the warm rays of the sun, and there would follow two great movements; the warm tropical waters flowing toward the poles, and the waters of the polar regions toward the Equator.

6. *As Cold Water is Heavier than Warm Water*, the latter would leave the Equator as surface or upper currents, and the cold water would approach it as under currents. Under these circumstances, the directions of the currents would be north and south. Besides this, *the Water which is taken up from the Tropical Regions by Evaporation*, is replaced by water flowing from the direction of the poles.

7. Allowing the *Earth to Revolve on its Axis* from west to east, and, remembering that the motion of the

surface is most rapid at the Equator and diminishes toward the poles, you will observe that as the waters from the polar regions approach the Equator, they are unable to acquire the more rapid motion of that part of the earth; consequently, the *Water falls behind*, and presents the appearance of a current rushing from east to west, round and round on each side of the Equator; this is called the *Equatorial Current*.

8. The *Course of the Equatorial Current is changed* by the deep sea-slopes of the continents and islands. The eastern angle of South America is so situated that the Equatorial Current is divided at Cape St. Roque.

9. The *Northern Section* of the Equatorial Current here takes a north-westerly direction, enters the Gulf of Mexico between Cuba and Yucatan, and issues from it between Cuba and Florida, and then turns north-eastward, constituting the Gulf Stream.

10. While the *Equatorial Current* appears to seek a westerly direction, it actually moves with the earth eastward; and, although not fast enough to keep up with the unyielding land of the Equatorial regions, still, when transferred to those parts of the surface whose easterly motion is less rapid, the Equatorial Current retains sufficient of its actual easterly velocity imparted to it when near the Equator, to go ahead of those parts nearer the poles.

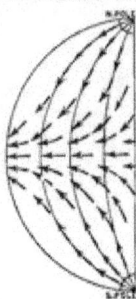

Boat Race Illustrating Currents of the Ocean. The Starting Points are shown by the Three Outline Figures on the Left.

11. *When you are on a Steamboat*, its motion causes the water, rocks, and trees near by to appear as if rushing past you in the opposite direction; even when you pass a boat which is sailing in the same direction with you, but less rapidly, it appears to move behind and away from you.

12. *In the Illustration above, the Steamboat represents* the land of the Equatorial regions; the small boat in which are two oarsmen, represents the water of those regions. Although both started together as shown in the left of the pic-

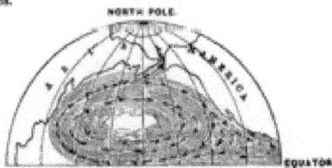

ture and moved in the same direction,—from west to east,—the swifter motion of the steamboat causes it to leave the oarsmen behind; consequently they appear to the people on the steamboat to move in the opposite direction,—from east to west.

13. The *Two Oarsmen represent* the Equatorial Current; they actually move *eastward*, but *apparently westward*.

14. Now compare the motion of the boat containing the two oarsmen with that of the boat containing but one, and it will readily be seen that the former goes ahead of the latter, and moves to the *east*; here, the two oarsmen represent the *Return Equatorial Current* flowing eastward, which in the North Atlantic is called the *Gulf Stream*, while the one oarsman represents the regions toward the poles, where the eastward motion of the Earth on its axis is slower than at the Equator.

15. The *Waters of the Equatorial Current* and the Gulf Stream are warmer than the other waters of the ocean, and have an important bearing upon the climate, productions, and inhabitants of the countries coming under their influence.

16. To the Gulf Stream *Europe is greatly Indebted* for its healthful climate, rich productions, and the general prosperity of its people.

17. The *Numerous Inlets from the Sea* which give to Western and Southern Europe an exceedingly extensive coast line, are peculiarly fitted for the distribution of the favorable influences of the Gulf Stream.

18. *Disconnect North and South America* by an extension westerly of the Caribbean Sea or the Gulf of Mexico, so that the Gulf Stream would flow into the Pacific, and the prosperity of Europe would be suddenly diminished; the *Mild and Genial Climate of the British Isles and France* would be exchanged for that of the bleak coasts of Labrador and Newfoundland, which lie between the same parallels.

19. In the same manner, the *Equatorial Current of the Pacific* continues westward until it reaches the islands east of Asia, where the northern part of the current is turned north-eastward to higher latitudes, where its easterly velocity predominates.

The Equatorial and Japan Currents of the Pacific Ocean.

20. *Under the Name of the Japan Current* it then flows eastward across the Pacific, until turned by the western side of North America, when, following the direction of the coast, it meets the Equatorial Tropical Current.

21. Therefore, the *General Plan of the Equatorial Current* is a flow round and round in ellipses, westward on or near the Equator; turning to the north in the Northern Hemisphere, and to the south in the Southern Hemisphere.

22. The *Equatorial Current flows in Deep Waters*, and its course is bent by the steep sides of the ocean's bed, about 100 miles from the coast line.

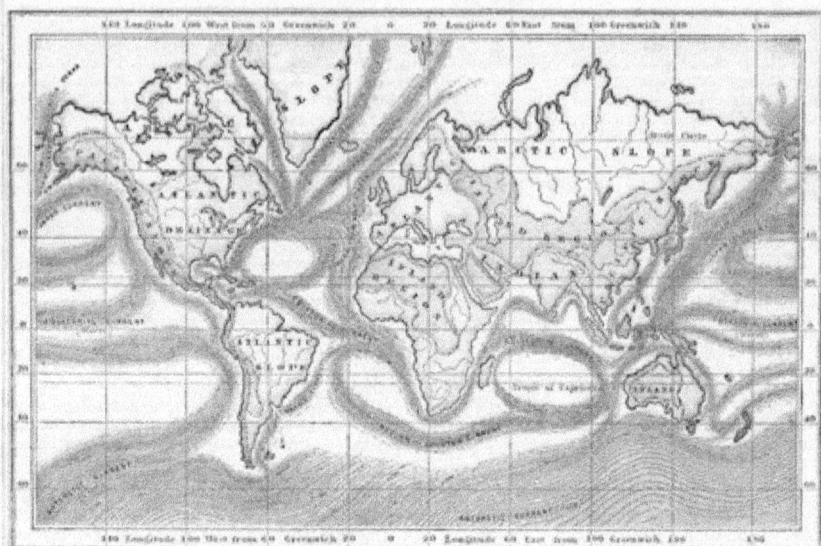

23. *From the Arctic to the Atlantic Ocean* two cold currents flow southwardly; one being west, the other, east of Greenland. These are called Arctic Currents; and, being unable to acquire the easterly velocity of those parts of the earth's surface which they pass on their way south, they are thrown to the west side of the ocean.

24. The *Arctic Currents carry* with them huge icebergs; many of which, as they meet the warm waters of the Gulf Stream off the coast of Newfoundland, become melted, and there deposit quantities of gravel, sand, and stones, transported from more northern lands.

25. *These Masses contribute* to the formation of the famous banks or shoals of that region.

26. Here, also, the *Cold Currents of the Atmosphere from* ture of summer as far north as the Banks of Newfoundland Evaporation from its warm waters is very rapid, hence the dampness in the atmosphere of the Atlantic States when easterly winds prevail.

30. The *Gulf Stream*, on reaching the British Islands, is divided; one part entering the Arctic Ocean, while the other is turned southward along the south-western coasts of Europe, where *its effect upon Atmosphere and Climate* is visible in the fertile vineyards and beautiful landscapes of that section.

31. The *Average Velocity of the Gulf Stream* is one and a half miles an hour; off the coast of Florida it is most rapid, being from three to five miles an hour. In the Pacific Ocean the Equatorial Current moves at the rate of about three miles, and, in the Indian Ocean, of two and a quarter miles an hour.

35. The *Equatorial Current of the Indian Ocean* connects with that of the Atlantic by a westerly current which doubles Cape Good Hope, called the Cape Current, in which vessels sail that are bound westward. South of the Cape Current is the return or counter current, in which vessels sail that are bound eastward.

36. *Vessels Navigating the Pacific*, between North America and Asia, sail westward in the Equatorial Current, and eastward in the return flow, called the Japan Current.

37. These two currents together form a great ellipse; its southern side being the Equatorial Current, and its northern side, the Japan Current.

38. *From the Japan Current, a Stream of Warm Water flows Northward* through Behring Strait; this, with a similar current from the Gulf Stream, tends to moderate the cold of the Arctic region, and to balance the cold currents flowing south on both sides of Greenland.

39. The *Climate of a Country depends chiefly* upon its latitude and elevation. It is also affected by the ocean and its currents.

40. The *General Flow of the Ocean Currents*,—westward in the tropical, and eastward in the temperate regions,—coincides with the atmospheric movements. In the tropics the winds blow to the west, and are called Trade Winds; in the Temperate Zones they blow to the east, and are called Return Trade Winds.

41. The *Temperature of the Atmosphere* is regulated by winds, or currents of air; while that of the ocean is regulated by currents of water.

42. Besides the *Great Benefits of the Ocean* already mentioned, there is another, in its myriads of fishes, which afford food and luxury to man; and, it is an interesting fact that the best fish are found in the cold currents, near the coasts.

43. The *Observing Learner cannot fail to see* that the ocean, which to the thoughtless appears as a great waste, is vast in its benefits; for it provides man with rain and streams to bring forth grass, fruit, and grain; tempers climates; bears his ships from nation to nation, and furnishes its living creatures as food for his table.

44. *Were the Warm Currents not turned toward the Poles*, the polar waters, now open, would be continually covered with vast fields of ice; hence, the coasts of America, extending far north and south, and turning the currents in their various directions, were thus formed according to a wise design.

45. The *Unceasing Activity of the Waters* of the ocean contributes largely to the benefit of all vegetable and animal

Section of a Hill, whence issues a Spring.
A. Loose Earth or Broken Rock through which the Rain sinks.
C. Solid Rock or Hard Clay not penetrated by Water.
S. Seam or Stratum in which the Water flows.

Section XII.

EVAPORATION, SPRINGS, AND WELLS.

1. *To the Ocean*, although salt, do we owe all the fresh water of the land. It is the source whence all springs, rivers, and lakes are supplied. The ocean and its streams of fresh water throughout the land, resemble the heart and veins by which the life of an animal body is sustained.

2. The *system by which the Land receives from the Salt Ocean a Bounteous Supply of Fresh Water*, is remarkable, as much for its completeness, as for the benefits which it imparts.

3. *All is the effect of* the combined action of heat, cold, and air. Heat lightens the water, that the air may lift it from the ocean; the winds carry it in the form of vapor over the land; the cold makes the vapor heavier than the air, and then it falls in the form of rain, snow, hail, and dew.

4. *The Rain that falls upon the Ground* serves to water the fields, and to fill lakes, rivers, ponds, and cisterns, for man's use. A part of it sinks into the ground, and forms subterranean streams or reservoirs; other portions are evaporated, and they again return, either to the land or to the ocean.

5. *Without Evaporation*, there would be no rain or dew, trees or grass; the whole land surface of the earth would be parched and barren.

Section of the Ground or Rock, showing how Wells are supplied.

A. The Part through which the Rain Water percolates.
C. Rock or Clay impervious to Water.
B. Seam or Stratum in which the Water passes.

10. *Wells are supplied* with water from the stratum in which it rests or flows, or with that which finds its way into them, through the crevices of the rock.

11. *Springs may be supplied by* rain or snow that falls on elevated ground several miles distant.

12. *After a Dry Season,* the flow from most springs becomes diminished, and sometimes ceases, until replenished by rain. There are, however, some springs whose discharge is uniform throughout the year; these are supplied from subterranean reservoirs, too extensive to be materially affected by ordinary droughts.

13. The *Quality of Spring Water* depends upon the materials composing the rocks or soil through which it flows. That which issues from sand-stone rock is softer and purer than that flowing through lime-stone strata.

14. *Intermittent Springs* are those which flow, and cease to flow, during alternate periods throughout the year.

15. *Mineral Waters* are those which possess medicinal qualities, owing to certain mineral substances which they hold in solution. There are, also, springs of salt water.

16. *Mineral Waters* are used for purposes of drinking and bathing. Mineral springs are numerous in the United States; *the most celebrated* are those of Saratoga and Virginia. They abound, also, in England, France, and Germany.

17. The *Strata at the Sides of the Continents* being inclined to the ocean, many subterranean streams empty into it, through its bed. In some instances, these streams are forced

20. *Boiling or Hot Springs* may be illustrated by a kettle partly filled with water, and placed upon a hot stove; the kettle representing the subterranean cavern, and the stove, the heated rocks of the earth's interior. The steam, if prevented from escaping at the top, presses upon the hot water below it, and forces it out through the spout, as shown in the illustration above. When the water in these caverns is long boiled and exposed to great heat, steam may be so suddenly generated as to produce explosion; this may account for the geysers (ghī'zers), or fountains of boiling water.

21. *Geysers* are of various dimensions; some are constantly boiling, others boil up only at intervals, with loud explosions.

22. The *Most Celebrated Geyser Regions* are in Iceland, California, and near the headwaters of the Yellowstone and Madison Rivers in the United States. The geyser region of the Yellowstone and Madison Rivers is more wonderful than any other

Artesian Wells:—A, A, A, Strata impervious to Water;—B, B, Seams or Strata in which Subterranean Streams flow;—C, Subterranean Reservoir filled with Water;—D, D, Borings in the Ground or Rock, called Artesian Wells.

25. *But how Complete is the Design in* this particular, also! The land is laid out by the hand of Providence, in channels and hollows, with streams, lakes, and reservoirs of water, on the ground, and under the ground, according to the plan which best contributes to the benefit of mankind.

26. *By Boring or Drilling into the Earth,* streams are met with at different depths, which are separated from each other by strata of rock; through the opening made, the water will rush upward as through a pipe, and rise like a fountain.

27. *These Openings, or Borings, are called* Artesian wells, from Artesium, now Artois, a province of France, where they have long been in use.

28. *In many Places Water has been thus obtained* in quantities sufficient for the working of heavy machinery.

29. *In Dry and Desert Regions,* even in Sahara, Artesian wells have been successfully sunk.

30. *Some Artesian Wells have been sunk to Depths* exceeding 2,000 feet, 84°; whence issues warm water; its temperature being derived from the internal heat of the earth.

31. *In Wurtemburg,* this water is introduced into pipes, for the heating of buildings, in winter; and by this means alone, the uniform temperature of 47° is maintained, while the temperature without is at zero.

32. *At Paris,* where the mean temperature at the surface, is 51°, the water of an Artesian well which is 1800 feet deep, has a constant temperature of 82°.

33. *At St. Louis,* the mean difference in temperature between the water obtained from an Artesian well, 1,500 feet deep, and that at the surface, is eighteen degrees.

34. *At Charleston, S. C.,* the temperature of the water at the surface averages 68°; at the depth of 500 feet it is 73°; at 1,000 feet, 84°; the average rate of increase of heat being about one degree for every 52 feet in depth.

35. *Many such Wells,* in New York, Pennsylvania, Virginia, and Ohio are famous for the quantities of salt and rock oil, or petroleum, obtained from them.

36. *Petroleum has been collected,* for centuries, in Birmah, Farther India, where it has been extensively used for producing artificial light; so, also, in northern Italy.

Section XIII.
Rivers; their Sources.

1. *Rivers are Formed from Springs,* or from rains that fail to penetrate the ground.

2. *They commence as* little streams, called *Rills, or Rivulets,* through which a child can wade, or over which he can step.

3. *Always seeking the Lowlands,* these rivulets meet other streams; and, enlarging as they go, soon become rivers.

4. *Like a Dove set free,* rivers seek their former home,—the ocean,—whether it be through extended plains, winding valleys, or mountain gaps. "Unto the place whence the rivers come, thither they return again." The dove seeks its home from a natural instinct; rivers seek the sea in obedience to the law of gravitation.

5. *Some Rivers rise in Regions of great elevation,* and at great distances from their mouths.

6. The *Sources of the Amazon* are far up the Andes; and, although they are within 100 miles of the Pacific, that river flows into the Atlantic, over a distance of about 4,000 miles.

7. *Rivers are useful as* great drains of the land; running off the surplus rain water into the ocean, and removing impurities from the surface of the ground. They also afford means of easy internal communication.

8. The *Courses of Rivers* are various, and are always governed by the slopes of the lowlands. Therefore, the general slopes of continents or countries can be determined from a common map, by the directions in which the rivers flow.

9. *We Observe that nearly all the Rivers of South America* flow in an easterly direction; hence, we know that the land east of the Andes, slopes towards the Atlantic.

Name the principal rivers of South America.

10. The *Rivers of Northern Asia and Europe* flow into the Arctic; hence, we know that from the Altai Mountains, the land presents a northern slope.

Mention the principal of these rivers. Mention the rivers of Eastern Asia, and the directions in which they flow. What is the slope of the land?
In what direction does the land of Southern Asia slope? Mention the largest rivers of the southern slope.

11. The land west of the Rocky Mountains slopes in what general direction? In what direction does the land of the United States, east of the Appalachian chain, slope? How do you ascertain this? The rivers of the United States, between the Rocky and Appalachian chains, flow into what river?

12. Mention the largest rivers on the western slope of the Mississippi basin; on the eastern slope.

What is the slope of the land of the Gulf States? Name the rivers of the southern slope.

How does the land slope in the region of Hudson Bay? Of the Baltic Sea? Of Western Africa?

13. The *Sources of Rivers always occupy Higher Ground* than do their mouths; many rivers, like the Ganges, have their sources several thousand feet above the level of their mouths, and owe their commencement to the melting snows of lofty mountains; consequently, their course to the sea is, at first, over very steep beds, or over a series of declivities, down which they plunge, producing rapids, cascades, and waterfalls. They approach their termination over beds less inclined, and comparatively level.

14. *Some Rivers*, like the Indus and Brahmaputra, flow for many miles on plateaus; others flow over beds of slight inclination from their sources to their mouths, and have no definite watershed. A boat may safely descend the Amazon River from the foot of the Andes Mountains to the Atlantic Ocean.

15. The *Waters of the Amazon are supplied* mainly by the excessive rains for which the Equatorial regions of South America are celebrated.

16. The *Upper Course of a River* commences at the watershed and continues over that part of its bed which is the most inclined: in this part, waterfalls and rapids are chiefly found.

17. The *Lower Course of a River* is toward its mouth; its bed is quite or almost level.

18. *By means of the Dissolving and Carrying Powers of Water*, the surface of the lowlands has received its comparative evenness.

19. The *Most Important River in North America* is the Mississippi. Its source is Itasca Lake, in the northern part of Minnesota, and is elevated nearly 1,700 feet above the level of the sea. Its general course is southward, and its total length about 3,000 miles.

20. The Mississippi is navigable by steamboats to the Falls of St. Anthony, 2,200 miles from the Gulf of Mexico; above the falls, it is also navigable for a considerable distance.

21. The *Mississippi and Ohio Rivers* constitute a line of communication between New Orleans and Pittsburg, of about 2,300 miles in length.

22. On the *Missouri River* steamboat navigation has reached to the foot of the Rocky Mountains, in the western part of Montana, a distance from the Gulf of Mexico of 4,000 miles.

23. The *Illinois River* is navigable for steamboats as far as La Salle, which is connected by a canal with Chicago, rendering navigation complete between Lake Michigan and the Mississippi, or between the Great Lakes and the Gulf of Mexico.

24. That the *Course of a River should not be in a Direct Line* to the sea, was wisely ordered by the Creator; for its various windings render the descent more gradual, and the current less rapid and destructive. Besides this, the winding course of a river increases the area of drainage, and the facilities for the progress of civilization and trade.

25. The *Distance from Cairo, Illinois, to New Orleans*, by the Mississippi River, is 1,178 miles. If there were no bends or windings in that river, the distance between these two places would be 700 miles less, but the force and destructive-

Passage worn through the Rocks on the Southern Coast of Norway.

28. *A River System* is composed of a river and its tributaries; thus resembling a great vine with its branches spread in all directions.

29. *A River Basin* comprises all the land that is drained by a river and its tributaries. In the lowest part of the valley flows the principal stream.

30. The *Basin of the Amazon* covers an area of more than 2,000,000 square miles; that of the Mississippi, about 1,000,000 square miles.

31. *A Watershed* is the ridge of land which surrounds a river basin and casts the water in different directions.

32. The *Watershed of Rivers flowing down opposite Sides of a Mountain Range*, is that part of the range which is elevated above the sources of those streams.

33. *A River Bed* is the ground over which the water flows. The channel is the deepest part of a river. The right bank is on your right hand as you sail down the stream; the left bank is on your left hand.

34. In many instances, *Springs but a few Rods distant*

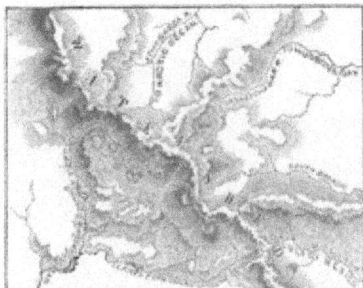

Watershed and Head-waters of Four Great River Basins in North America.

36. The *Head-waters of the Missouri and Clarke's Rivers*, in the Rocky Mountains, are almost together; yet the waters of one, by way of the Missouri and Mississippi Rivers, enter the Gulf of Mexico, and thence into the Atlantic; while the waters of the other, empty into the Columbia River and find their way into the Pacific.

37. *A Northern Tributary of the Columbia River* has its head-waters very near those of the Saskatchewan and Athabasca Rivers.

Where do these rivers rise? Into what does the Columbia River flow? The Missouri? The Mississippi? The Saskatchewan? The Athabasca? Which are on the eastern slope of the Rocky Mountains? On the western slope?

38. *A House may be so located* upon the ridge which forms a watershed, that the rain falling upon one slope of its roof, may eventually find its way to one ocean, and that falling upon the opposite slope, to another ocean.

39. *Asia differs greatly from North America*, in this respect. The river basins of the Indian, Pacific, and Arctic slopes, are so disposed that the head-waters of their rivers are separated from each other by vast plateaus.

40. *Some Rivers do not empty into the Ocean*, but into an inland sea, or lake, as those of the Caspian Sea basin and the great basin of Utah. The river Jordan which flows into the Dead Sea belongs to this class.

41. *Some Rivers of Africa* disappear in the sands of the desert, and others are partly subterranean. These enter caverns, channels, or loose strata below the surface.

42. *Oceanic Rivers* are those whose waters reach the ocean, directly or indirectly; as the Amazon, Ohio, Danube, and Connecticut.

43. *Continental Rivers* are those of inland regions, whose waters do not reach the ocean; as the Volga and Ural.

44. *Many Rivers which have Rapid Currents* bear along with them alluvial washings from the land, and deposit them at their mouths, forming deltas.

45. The *Mississippi* and its tributaries are constantly transporting mud, logs, and stones, from the land of about twenty States and Territories, and depositing them in the valley of the Mississippi and at its delta.

46. *Borings have been made*, north of New Orleans, to the depth of 600 feet without reaching the bottom of the drifted mass; and, judging from the amount annually brought down by the Mississippi, it is estimated that the formation of land by its deposits, has already occupied more than 100,000 years. Hence, the land is constantly encroaching upon the Gulf of Mexico.

47. *This is also remarkable* in the Ganges, Nile, and Rhine.

48. Accordingly, the mountains and hills on the globe are being gradually diminished in height, and the land surface of the earth gradually extended.

49. The *Streams rushing down the Mountain Sides*, are constantly carrying new soil to increase the fertility of the plains below. On their way down they turn the wheels of numerous mills and manufactories; and, by means of reservoirs and pipes, cities are abundantly supplied with fresh water.

Waterfall—Saw Mill.

50. The *Water of a River* is high or low, according as the season is rainy or dry.

51. *Many Rivers, like the Mississippi*, become full, sometimes to overflowing, by the melting of the snow at the approach of spring; but, during the summer months, the water is comparatively low.

52. *Other Rivers, like the Nile*, receive the tropical rains and rise periodically.

53. The *Sources of some Rivers*, like the Mackenzie and those of Siberia, are affected by the spring thaw, while their mouths, far northward, remain covered with ice; causing extensive overflows, by which stones, masses of earth, trees, and ice, are carried far across the land.

54. *Rivers opening into the Ocean* receive sea-water which is forced into them by the tides and winds; thus increasing their importance for purposes of navigation.

55. This is remarkable chiefly with rivers which open toward the east and south, owing to the westward movement of the tide.

56. The *United States and Europe* owe much of their greatness to their rivers, canals, and railroads, which intersect all their important parts.

57. *All that part of Europe lying West of the Black Sea*, is traversed by rivers which rise in the same region, and flow in all directions; while Asia and Africa contain immense tracts not crossed by a single river.

58. The *Importance of Rivers* to the development of mankind is manifested by the numerous villages and cities which line their banks; thus resembling the vine, whose value is indicated by the clusters of grapes hanging upon its branches.

59. Although South America is still in a backward state of development, its vast rivers and fertile plains promise it, in the future, a high rank among the divisions of the earth.

A Sectional View of the Great Lakes and the St. Lawrence River, looking North.

Section XIV.

LAKES; THEIR ELEVATIONS AND DEPTHS.

1. *Lakes* are collections of water in hollows of the land, of such a depth that their outlets cannot completely drain them.

2. *There are Four Classes of Lakes:*

3. The *First Class* has no streams which serve either as inlets or outlets.

4. The *Second Class* differs from the first in having an outlet; both classes are supplied by springs which burst forth from the bed of the lake.

5. The *Lakes of the Second Class* are generally situated on great elevations, and, in many instances, form the sources of rivers.

6. The *Third Class* both receives and discharges its waters by means of streams. Most lakes belong to this class.

7. The *Fourth Class* includes those lakes which receive streams of water, but have no visible outlet. They belong to continental or inland basins, and are numerous in Asia. These lakes are kept from overflowing their banks by means of evaporation.

8. *Many Depressions of the Land Surface* would contain lakes, but for the effect of evaporation.

9. *Nearly all Lakes* are supplied by streams which empty into them, and by springs rising from the bottom and sides.

10. *Some Lakes in Mountainous Regions* are supplied from the melting snow of the surrounding peaks.

11. *Lakes* occur in highlands and lowlands. Some are elevated several thousand feet above the sea level, while others are depressed below it.

12. The *Most Elevated Lake in the World*, is Lake Sir-i-kol, which is situated on the mountains in the western part of the Chinese Empire. It is about 15,000 feet above the level of the sea. (*See Illustration on page 17.*)

13. *Lake Titicaca*, between Peru and Bolivia, is over 12,000 feet above the level of the ocean. Its area is more than 2,000 square miles, and its depth is equal to that of Lake Ontario.

14. The *Dead Sea*, properly a lake, is more than 1,300 feet below the sea level. It is the greatest depression of the kind on the globe. This famous lake, whose formation resulted from the catastrophe which destroyed the cities of Sodom and Gomorrah, about 1,900 B. C., contains a far greater portion of salt than do other salt lakes; the water being so impregnated with it, that even heavy bodies float buoyantly. Asphaltum, in large quantities, and sulphur, are found on its banks.

15. The *Waters of most Lakes are Fresh;* but those having no outlet are usually salt. This is because all streams receive from the land through which they flow, small quantities of salt, which the waters hold in solution until it reaches the ocean, or another body of water having no outlet; here the salt is deposited.

16. The *Most Celebrated Salt Lakes* are the Caspian Sea, Aral Sea, and Dead Sea, and the great Salt Lake of Utah.

17. The *Basin of a Lake* comprises all the land drained by the streams which flow into the lake. It may be seen on a map, by passing a line around the sources of all its tributaries.

18. *Subterranean Lakes* are numerous. They are collections of rain water in caverns which are below the surface of the ground.

19. *Subterranean Lakes and Streams* frequently cause destructive inundations. The water and steam thrown up by volcanoes proceed from these lakes.

20. The *Island of Trinidad*, situated near the mouths of the Orinoco River, contains a lake three miles in circumference, that is famous for the quantities of pitch contained in its waters. This substance, like petroleum, is raised by the agency of subterranean fire.

21. *Lake Superior* is the largest body of fresh water on the globe. Its area is 32,000 square miles, and is equal to about three-fifths that of England.

22. Are the waters of the Great Lakes salt, or fresh? What river forms their outlet? In what direction does the St. Lawrence River flow? Mention the depth of each lake. Which is the deepest,—the shallowest? Which has the most elevated surface? Between what two lakes are the Falls of Niagara situated? From which does the water of the falls proceed?

In what part of the St. Lawrence are the Thousand Islands? The Rapids?

Section XV.

The Atmosphere;—The Winds.

1. The *Atmosphere* is a gaseous fluid which surrounds and rests upon the earth.

2. *It is as necessary to Life*, as are water and food; neither plants nor animals could exist without it.

3. *Air consists of Two Gases*, oxygen and nitrogen, mixed together.

4. *The Ingredient of the Air which sustains Animal Life,* is oxygen; but, should these two gases be separated, the result would be instant death.

5. The *Air, like Wholesome Food*, is necessarily composed of both nutritious and innutritious substances.

6. *Oxygen forms* about one-fourth of the air; nitrogen, three-fourths.

7. The *Weight of the Atmosphere* is about $\frac{1}{800}$ that of water.

8. *It is Heaviest* at the surface of the earth, and diminishes in density, according to the distance above the surface.

9. *On the Tops of the Highest Mountains*, the air is so thin that man cannot breathe there.

10. The *Atmosphere extends upward*, to a distance, it is supposed, of about fifty miles.

11. *Winds* are currents or movements of the air, caused by the different degrees of temperature to which the air is subjected, and by the revolution of the earth upon its axis.

12. The *Air is Warmed*, partly by the passage through it of the sun's rays, but mostly by the radiation of the sun's heat from the earth's surface; consequently, the warmest part of the atmosphere is that which is in contact with the surface of the earth.

13. The *Heat and Density of the Atmosphere* diminish according as the elevation is increased; this has been observed by travelers who have ascended high mountains, and by aeronauts in their balloon ascensions.

14. *As that part of the Atmosphere is Warmest which is nearest the Surface*, the upper and surrounding cold air presses down and replaces the warm and light air, which rises to more elevated regions.

15. *A Balloon ascends because it* is filled with a gas that is lighter than common air. When the gas is allowed to escape, the surrounding air rushes in and causes the balloon to descend.

16. The *Two General Movements of the Air* are from the Equator to the Poles, and from the Poles to the Equator.

17. *As the Cool and Heavy Winds press toward the Equator*, they are unable to keep up with the *eastward* motion of the Equatorial regions of the earth; and, by falling behind, they appear as a current of air moving *westward*.

(For further explanation, see page 23, paragraphs 6 and 7.)

18. *A Current of Water* receives the name of the direction *toward* which it flows; but a current of air, that *from* which it moves. Therefore, a *westerly* current of water and an *east* wind are in the same direction.

19. *Changes in the Course of the Winds* are caused by various bodies of land, and by high mountain ranges.

20. *Where Two Winds from Different Directions meet*, they counteract each other's force, and cause calms; hence, there are Equatorial Calms, Calms of Cancer, Calms of Capricorn, and Polar Calms.

21. The *Trade Winds* of the Northern Hemisphere blowing from the north-east, and those of the Southern Hemisphere blowing from the south-east, meet near the Equator, and neutralize each other; thus causing calms in that region around the earth.

22. The *Winds* then rise to a greater elevation and tend toward the North and South Poles, moving over the tropical regions as upper currents.

Chart showing the Directions of the Winds.

23. When they reach the temperate latitudes, they have become so cool and heavy on account of their elevation, that they descend to the surface, and blow from the south-west in the Northern Hemisphere, and from the north-west in the Southern Hemisphere. These are called *the Return Trades or Passage winds.* (See page 24, paragraphs 10 to 14 inclusive.)

24. The *General Direction of the Winds in the Tropical Regions* is toward the west. These winds contribute to the westward flow of the Equatorial Current.

25. *In the Temperate Regions* there is a like correspondence between the Return Currents of the ocean and the Return Trade Winds; their motion being toward the east.

26. *A Voyage from the United States to England*, in a sailing vessel, is made several days shorter by the aid of these winds and the Gulf Stream, than that from England to the United States. The time made by steamers from New York to Liverpool, is between nine and twelve days; but, returning, they require from two to four days longer.

27. The *Prevailing South-west Winds of the North Temperate Zone*, passing over the warm waters of the Gulf Stream, contribute largely to the advantages of Western and Southern Europe in climate, productions, and general development.

28. *If the Earth revolved on its Axis in the Opposite Direction*—from east to west—in what direction would the Trade Winds and the Equatorial Current move? If the earth did not revolve on its axis, what would become of the ocean currents and the winds?

29. The *Plan of the Winds*, like that of the ocean currents, is such that a constant circulation of air is maintained between the Eastern and Western Hemispheres, round and round the globe, and between the Northern and Southern Hemispheres, from the burning zone of the Tropics to the frozen regions of the Poles. "The wind goeth toward the south, and turneth about unto the north; it whirleth about continually."

Sea Breeze. A View on the Sea Coast. From Morning until Evening the Air which is over the Sea is Heavier than that over the Land; consequently, the Wind blows all Day from the Sea.

30. The *Plan of Differences in Nature*, producing contact, opposition, and variety, is beneficial to mankind.

31. *It is recognized* in the light of day and the darkness of night, in land and water, sunshine and rain, in the variety of productions, and in the diverse pursuits of people.

32. The *Wisdom of this Plan* appears not only in the existence of such differences, but in their coöperation and unity.

33. *Sea Coasts and Islands* enjoy a more even temperature throughout the year than inland districts, because the ocean does not change its temperature, either in summer or winter, so readily as the land.

34. *Winds* which blow over the sea are generally not so cold in winter, nor so warm in summer, as those blowing over the land.

35. *Land near the Sea is Warmer during the Day* than the neighboring water. Sea air is then cooler and heavier than the air of the land; hence, the wind blows all day from the sea, and is called a *sea breeze*.

36. *After Sundown*, as the land becomes cooler than the water, the air rushes back from the land, and is called a *land breeze*.

Land Breeze. At Night, the Air which is over the Land becomes Heavier than that over the Water, causing the Wind to blow all Night from the Land.

37. *Land and Sea Breezes* are winds which blow alternately from the land and sea.

38. They occur on coasts and in islands, especially in the tropical regions; also on the shores of large lakes.

39. *In the Northern Hemisphere*, a north wind is cold, and a south wind, warm; in the Southern Hemisphere, the north wind is warm, and the south, cold.

Section XVI.
Moisture in the Atmosphere.

1. By heat, *Water is Expanded* and made lighter than the air.

2. The *Water then rises* in the form of vapor, and is carried away by the winds.

3. *Vapor when Influenced by a Cool Temperature* becomes condensed, and returns to the surface of the earth in the form of rain, snow, and dew.

4. The *Motive Power of the Steam Engine* is due to the property which water possesses of being easily expanded by heat and condensed by cold, thus forming a vacuum.

5. *As the Air becomes Warm*, its capacity of holding moisture increases, and as the temperature falls that capacity diminishes. This difference between the temperature of the day and that of the night, causes dew to appear upon the grass and flowers, that they may be refreshed in the absence of rain.

6. *Trees and Plants* obtain much of their nourishment from the moisture in the air which is condensed by means of their leaves.

7. *Vapor is not always Visible*, because it is spread out in the atmosphere, like the moisture that is exhaled in breathing. A pitcher of cold water placed in a warm room condenses vapor, which appears on the surface in the form of drops.

8. *Evaporation increases* with the warmth and dryness of the atmosphere; hence, the amount of rain is greatest in the tropical regions, and diminishes toward the poles.

9. *Evaporation modifies Temperature.* Without evaporation, the surface of the ocean would become hotter and hotter by the influence of the sun, and would therefore greatly intensify the heat of the atmosphere in contact with it. But not thus defective are nature's laws.

10. *As Water becomes Heated at the Surface*, it gives place to cooler portions beneath, by rising, in the form of vapor, into upper and cooler regions of the atmosphere.

11. *By the Action of the Waves*, lower and cooler portions of the water are brought up to the surface to reduce its temperature.

12. *By these Movements of the Water*, the surface of the ocean is prevented from attaining a degree of heat so great as to prove detrimental to the comfort and interests of mankind.

13. *On the Land*, these two movements do not occur. Its heated surface cannot rise in the air as water does by the process of evaporation; neither are cool portions of the ground brought constantly up to reduce the temperature of the surface; hence, the land becomes more heated by the sun's rays than the water does.

14. *In Summer*, the land freely imparts its heat to the atmosphere near it and makes that season hot, perhaps oppressive; but when winter comes, the land has not saved enough heat to keep off the severity of the cold. It is owing to this process of radiation that in some inland places, hot and sultry days are followed by chilly and disagreeable nights, and that the deposition of dew is greater on land than on water.

View from the Catskill Mountain House, New York, looking East. The Hudson River appears in the Distance.
Names of the Classes of Clouds :—1. Cirrus ; 2. Stratus ; 3. Cumulus ; 4. Nimbus.

15. *In Winter*, the continental climate is colder than the oceanic, because the land parts with its heat by radiation more readily than does the water.

16. *St. Petersburg and the Faroe Islands* are nearly in the same latitude : the climate of the former is *continental* ; of the latter, *oceanic*. Which is the warmer in summer? Which is the colder in winter?

17. The *Summer of St. Petersburg* averages seven degrees warmer than that of the Faroe Islands, north-west of Scotland; while the winter of the former is twenty-two degrees colder than that of the latter.

18. *Clouds* are collections of visible vapor suspended in the atmosphere, at altitudes ranging from one to five miles.

19. *Fog* is a like collection nearer the earth's surface.

20. *Vapor* consists of particles of water so fine and light that they float in the air like dust.

21. *There are Four Classes of Clouds :*

22. The *Cirrus*, which is the highest cloud we see, is of a light feathery form ; and, on account of its elevation, its vapors probably exist in light particles of snow.

23. The *Stratus* exists generally in the night and in winter; it is formed by the cooling and consequent settling down of the higher clouds, which appear in horizontal bands.

24. The *Cumulus* is the summer-day cloud which forms at sunrise by the gathering together of the night mists.

25. The *Nimbus* is the heavy, dark cloud from which rain falls.

26. *When Clouds pass into the Atmosphere* which surrounds the cold summits of the mountains, their vapors become condensed, and fall in the form of rain and snow, which supply springs, streams, and lakes of elevated regions.

27. *Rain falls from Clouds* at different elevations ; in mountainous districts heavy showers sometimes fall upon the low ground, while persons on a mountain behold a clear sky above them and black clouds below them.

28. *If there were no Mountains* on the globe, the clouds would pass over the land without depositing an amount of rain sufficient for the preservation of vegetable and animal life.

29. The *Harmony which exists* between the influence of the mountains and the movements of the clouds, produces results necessary to the development of the earth and to the well-being of man. Is this harmony the result of accident, or is it in accordance with the wise design of the Creator?

30. *Rain* is caused by vapor entering a cool atmosphere and becoming condensed ; it then falls to the earth in drops.

31. *If Rain, in its Descent, passes through a Current of Air* sufficiently cold to freeze the drops, hail is produced.

32. *If Vapor becomes Frozen* while its particles are light, it falls to the earth in the form of snow.

33. *In North America*, snow is seldom seen to fall south of the parallel of latitude 30°—that which passes over New Orleans.

34. *In the Hot Zone of South America*, however, it remains throughout the year on all mountain peaks above the elevation of 16,000 feet.

35. *Snow is a Non-conductor of Heat*; it consequently prevents radiation of heat from the ground covered by it, and protects roots, vegetables, and seeds from the intense cold of winter.

36. *Rain is distributed* over the land by the agency of winds.

37. The *Chief Source of Supply* is the ocean; although from every lake, pond, and stream, there arises moisture which returns to refresh vegetation.

38. The *Greatest Amount of Rain* falls within a belt around the earth, near the Equator. This is because the Trade Winds here come in contact with each other and carry the vapors with which they are heavily charged, up into a cool atmosphere which condenses them.

39. *On the Continents*, the greatest amount of rain falls near the sea coasts and upon the mountainous regions in the interior.

40. *On the Western Continent*, the greatest amount of rain falls in South America—on its eastern coast, and the eastern slope of the Andes Mountains.

41. *Ranges of Mountains*, like the Andes, whose tops are perpetually covered with snow, cause vast quantities of rain to fall on the windward side or slope, while in some places on the opposite slope, rain is almost or wholly unknown.

42. The *Desert of Atacama* (ah-tah-kah'mah) is situated west of the Andes, and lies partly in Peru and partly in Bolivia. The east winds are deprived of their moisture before passing the mountains, and continuing westward, prevent the vapors of the Pacific from reaching that arid region. (*See Sec. IX., par. 14.*)

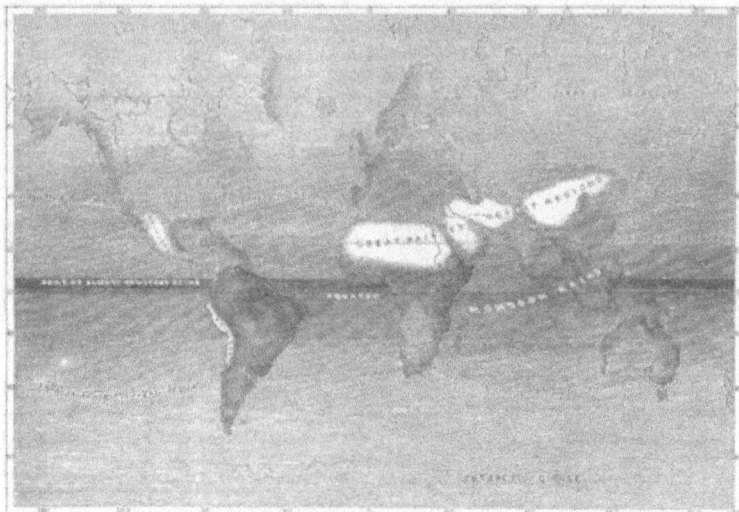

Rain Chart:—The Quantity of Rain which falls at any Place is indicated on this Chart by the Depth of the Shading; the Darker the Shading, so much Greater is the Amount of Rain.

43. The *Rainless Region of South America* lies west of the Andes, and in the track of the *South-east Trade Winds.*

44. *South of the Desert of Atacama* is Chili, which, although lying on the west of the Andes, receives copious rains; these are brought by the *Return Trade Winds*, which blow over the Pacific from the north-west.

45. The *North-western Coast of South America* is within the zone of almost constant rains.

46. The *Trade Winds* blow from the east and deposit their rain mostly on the eastern coasts of continents and islands, and on the eastern slopes of high mountain ranges.

47. The *Return Trades* deposit their rain chiefly on western coasts and slopes.

48. *In North America*, rain is most abundant on its western side, and around the Gulf of Mexico.

49. The *West Indies are Noted* for the heavy rains which fall there; these rains proceed from the vapors supplied by the warm waters of the Gulf Stream.

50. The *British Islands*, together with the western and southern coasts of Europe, are supplied with rain from the vapors of the Atlantic Ocean, which are carried there by the prevailing west winds—the Return Trades; while on the plains of Russia and Siberia, the amount of rain is comparatively slight.

51. The *Rains of Africa*, like those of South America, are supplied by means of the Trade Winds; but while South America has its high mountain range on its western side, causing copious rains to fall upon vast plains eastward, the high mountains on the eastern side of Africa return much of the rain immediately into the Indian Ocean.

52. *Do the Trade Winds deposit more Rain* on the eastern, or the western sides of islands and mountains? On which coast of South America is rain most abundant? On which coast of Africa? On which side of the Andes Mountains? On which side of continents, islands, and mountain ranges do the *Return Trades* deposit most rain?

53. *Do Vapors rise mostly from Cold, or Warm Currents?* From what current do vapors come which supply the rivers of Western and Southern Europe? From what current are the rivers of the Pacific coast of North America supplied? (*See Chart on page 28.*)

54. *On which Coast of Greenland is Rain most Abundant?* On which side of Norway? France? Spain? Arabia? Australia? Hudson Bay?

55. *What Great River in Africa* flows through the rainless district? Whence does the Nile receive its waters?

56. *If no Ocean intervened between America and Europe,* the absence of rain alone would make Europe desolate.

57. The *Great Rainless Region of the Old World* includes the Great Desert of Africa and the deserts of Arabia, Persia, and Cobi.

58. *Their Condition is caused,* mainly, by their interior position, the comparative dryness of the winds, and the absence of lofty peaks that would act as condensers of the thin and scattered vapor which floats over them.

59. *In the New World,* the principal rainless districts are in Mexico and Central America, and in South America, on the western side of the Andes.

60. *In some Places where Rain seldom or never falls,* vegetation is sustained by frequent and heavy dews.

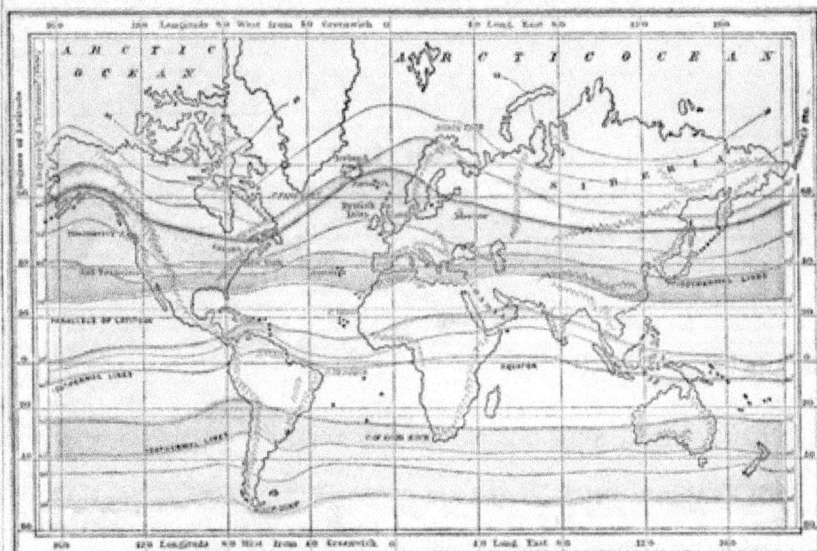

Chart showing Isothermal Zones and the Mean Annual Temperature of the Different Parts of the Earth's Surface.

Section XVII.

CLIMATE,—ISOTHERMAL LINES.

1. *Climate* is the condition of a place in relation, chiefly, to the temperature and moisture of the atmosphere.

2. *Isotherms, or Isothermal Lines,* are lines drawn on a chart through places of equal mean temperature.

3. *Mean Annual Temperature* is midway between the heat of summer and the cold of winter. In Cincinnati the mean temperature of summer is 73°, and of winter, 33°; the mean annual temperature is 53°, which is obtained thus:

$$\frac{73+33}{2} = 53.$$

4. *If the Temperature diminished uniformly from the Equator to the poles,* isothermal lines would correspond with parallels of latitude.

5. *Their Directions* are various, and indicate the influence upon climate, of ocean currents, winds, high mountains, frozen plains, and burning deserts.

6. Therefore, *the Hot, Cold, and Temperate Zones* of the earth are situated between isothermal lines, and not between parallels of latitude. These zones are called *Isothermal Zones.*

7. *Isothermal Lines have their Greatest Inclination* in the North Atlantic Ocean, and show that the north-west coasts of

the Old World possess warmer climates throughout the year than other parts of the land, at the same latitude.

8. *This is chiefly owing to the influence of the Gulf Stream,* which warms the prevailing south-west winds passing over it on their way toward the west coasts of Europe.

9. *Eastward from these Coasts,* the temperature gradually falls, as shown by the isotherms, on account of the cooling influence of the high mountains of Europe and Asia, and the frozen plains of Siberia.

10. *If the Waters of the Atlantic Imparted no Warmth* to the atmosphere, Newfoundland and Northern France, being between the same parallels of latitude, would have the same climate.

11. *Without the Influence of the Gulf Stream,* the now genial and productive climate of the British Isles would be similar to that of the cold and desolate regions of Labrador.

12. In reality, however, the *Centre of Great Britain,* at the latitude of 55°, has the same mean temperature as the eastern side of the United States, at the latitude of 40°.

13. The *Isotherm which passes through Newfoundland* extends north-eastward to the coast of Iceland, 15° nearer the North Pole.

14. The *Temperature of the coast of Norway* is the same as that of Central Labrador, although 20° of latitude lie between them. The influence of the Gulf Stream is felt upon the coasts of Spitzbergen and also upon the north coast of Nova Zembla.

15. The *Land of the Northern Hemisphere* may be divided into six climatic zones: The Torrid or Hottest, the Hot, Warm, Temperate, Cold, and Frigid or Coldest.

16. THE MEAN ANNUAL TEMPERATURE OF THE ZONES.

The Frigid Zone, below.............................	32° Faht.
The Cold Zone, between..........................	32° and 40° "
The Temperate Zone, between..................	40° and 60° "
The Warm Zone, between.......................	60° and 70° "
The Hot Zone, between...........................	70° and 80° "
The Torrid Zone is over..........................	80° "

17. The *Isotherm of 32° Fahr.* is the line of constantly frozen ground.

18. *Through what Parts of North America does the Isotherm of 32° pass?* Through what parts of Europe? Of Asia? What large bay in British America receives cold water from the Arctic Ocean? What effect has the temperature of the water of Hudson Bay upon the climate of the surrounding regions? What is the direction of the isotherms which pass over these regions?

19. *What Places are under the same Isotherm as New York?* What is their mean temperature?

What places are under the isotherm which passes over Panama? What is their mean temperature?

What places are under the isotherm which passes over Newfoundland?

20. *What Parts of the Northern Hemisphere are in the Hottest Zone?* The Hot Zone? The Warm Zone? The Temperate Zone? The Cold Zone? The Frigid Zone?

What lands of the Southern Hemisphere are in the Hottest Zone? The Hot Zone? The Warm Zone? The Temperate Zone?

Does any part of the two continents extend south of the line of constantly frozen ground? What part extends furthest south?

What is the mean annual temperature of Cape Horn?

21. The *Prevailing Winds of the United States and Europe* blow from the south-west; consequently, they are *Land Winds*, to the eastern parts of the United States and Europe, and cause *Excessive Climates* (see page 33, paragraph 34); while to the western coasts, they are *Sea Winds*, and produce that evenness of climate for which Western Europe and the Pacific coast of the United States are remarkable.

22. *If we Compare the Climate of New York* with that of San Francisco, the difference between oceanic and land climates will be obvious.

THE MEAN TEMPERATURE OF THE HOTTEST AND COLDEST MONTHS DURING THE YEAR, IN NEW YORK AND SAN FRANCISCO.

Hottest month in New York, 80° Fahr. ;—San Francisco, 58°.	
Coldest " " 25° " " 50°.	
Mean difference between sum-	
mer and winter............ 55° " " 8°.	

23. *While Snow usually lies in New York a great part of the Winter*, it rarely falls in San Francisco. The winter of San Francisco consists of a *Rainy Season*, which is caused by the cooling influence of the mountains upon the moisture of the sea winds. Its summer is known as the *Dry Season*.

24. The *Temperature of the East Coast of the United States* is further depressed by cold waters from the Arctic Currents, which here flow in a south-westerly direction between the Gulf Stream and the coast. It is therefore a counter current.

25. The *Valleys near the Coast of California* possess a more even and delightful climate than any other part of the world.

26. *In some parts of the Faroe Islands*, water never freezes, while in Yakontsk, a city of Siberia, which lies under the same parallel, the summers average 9° warmer, and the winters, 76° colder. The mean difference in temperature between summer and winter at the former place, is only 15°; at the latter, it is 100°.

27. *In which of the two Places* just mentioned is the climate excessive? Even? Continental? Oceanic?

28. *In the Azores and Madeira*—islands north-west of Africa,—the climate is that of eternal spring; flowers bloom there throughout the year in the open air, although those islands are between the same parallels as Philadelphia, Cincinnati, and St. Louis.

29. *Forests, Fertile Plains, and Parched Deserts* owe their respective conditions not only to their position on the globe, but also to the influence of ocean currents, the agency of winds, and the presence or absence of rain.

30. The *Isotherms of North America, Europe, and Asia* extend in the same general direction—south-eastward from their western sides; showing the mean temperature of their western coasts to be warmer than that of their eastern.

31. The *Climate of the Atlantic Coasts of Europe* corresponds with that of the Pacific Coast of North America.

32. Isothermal lines correspond more nearly with parallels of latitude in the Water Hemisphere than in the Land Hemisphere, showing the evenness of an oceanic climate.

33. *Compare the Climate of Vancouver's Island with that of Maine.* In the former, the summers are mild, and the frosts of short duration; while in the latter, the summers are hot, and the winters very severe, the snow lying on the ground from three to five months in the year.

34. *Traveling Eastwardly from the Pacific Coast* of North America on any parallel north of San Francisco, what change of temperature is observed? (See Isothermal Lines.)

35. *Sailing Due East from the Atlantic Coast*, what change? What part of the Pacific coast of North America has the same temperature as Newfoundland? Give the latitude of each of these two places. What is the average temperature?

What island on the Pacific coast of North America has the same temperature as New York? What is the latitude of each? Their mean temperature?

36. *What European Country* has a climate similar to that of California? Although North Cape is 11° farther north than Cape Farewell, its climate is no colder. Why?

What city in Russia has the same latitude as Glasgow? At which place is the winter more severe? Why?

37. *Why does the Climate of the West Indies* differ from that of Newfoundland?

Which is farther north—Canada, or Iceland? In which are the winters more severe? Why?

Which coast of the United States possesses the more even climate—the Atlantic, or Pacific? Why?

38. The *Climate of the Western Side of North America* and of Western Europe is more conducive to health than that of their eastern parts, on account of its greater evenness.

39. *If the Bed of the Atlantic should be elevated* and become dry land, what climates would be affected, and how?

If a range of high mountains extended along the west coast of Europe, what would be the effect upon the climate and rains of that division?

40. *Why is the Climate of the Atlantic Coast* of North America warmer in summer, and colder in winter, than that of the Pacific coast?

What effect have the Rocky Mountains upon the temperature of the westerly winds of the United States?

A Mountain Stream.

41. *Activity, Use, and Influence* are everywhere, from the mighty ocean and lofty mountains to the little stream that turns the miller's wheel and furnishes drink to cattle.

42. The *Common Garden Worm* opens channels in the ground through which the moisture enters to nourish the roots of plants, and otherwise assists man in preparing the soil.

43. The *Ocean*, although covering the greater part of the earth's surface, is not a vast waste, for it supplies the land with vegetation and an abundance of fresh water for the support of all life; and, as the modifier of climate, it exerts its essential influence upon the physical, intellectual, and moral conditions of mankind, and contributes largely to the prosperity of the nations of the earth.

The Earth in the form of a Globe.   The Earth in the form of a Cube.   The Earth in the form of a Cylinder.

44. *None can fail to recognize the Systems of Winds* and ocean currents as necessary to the life and well-being of the earth's inhabitants; and, herein, the wisdom of the plan by which the world was made in the form of a globe.

45. *If the World had been made in the Form of a Cube*, or of a cylinder, there would not be that harmony of action between diverse conditions of the earth's surface which now exists.

If the Earth were a Great Cube, would there be areas of different degrees of temperature as there are now? The same winds and ocean currents?

46. *Diversity in Climate and Productions* of the earth, and in the pursuits of individuals and nations, constitutes a wise provision of the Creator.

47. *All the Great Agents* by which the various conditions of the earth are so wonderfully sustained, are so adapted to each other, and act together so harmoniously, that if but one should neglect to act its part, mankind would suffer—perhaps perish.

48. *If the Process of Evaporation should be discontinued, what would be the effect upon vegetation, animals, and man? Or, if all winds should cease, where would all the rain fall?*

49. The *Southern Part of the United States* is admirably adapted to agriculture. Its peculiarities of soil and climate so harmonize with each other that the amount of cotton alone which is here produced, and upon which millions of the earth's inhabitants—on both continents—depend for clothing, comprises nearly seven-eighths of the entire yield of the world.

50. The *Rugged North-eastern Part of this Country* is provided with coal, iron, and mountain streams, which make it the great manufacturing region of the Union.

51. *If the Gulf and Atlantic States of the South* were mountainous, and the north-eastern States level, the cotton plant, sugar-cane, and rice would not grow either upon mountains of the south or cool plains of the north-east.

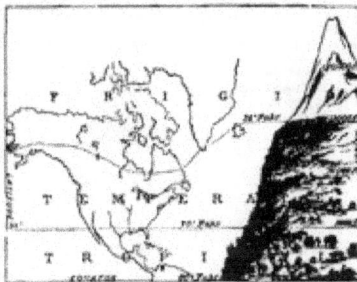

Chart, showing that Climates between the Equator and the North Pole correspond with those on the Sides of High Mountains at the Equator.

52. *Temperature so diminishes with Increase of Elevation* that various climates, with their characteristic productions, are found not only upon the earth's surface between the Equator and the Poles, but likewise upon the sides of high mountains between their base and summit.

53. *If we consider the Northern Hemisphere* and the side of a mountain which is situated under the Equator, to be divided each into three climatic zones, the Torrid Zone on the former would extend northward to about the parallel of 30°, and on the latter, upward to the elevation of about 5,000 feet; the Temperate Zone of the former would extend to about the Isotherm averaging 60° latitude, and on the latter, to the height of about 15,000 feet.

What part of the earth's surface and what part of a tropical mountain have a mean temperature of 80° Fahr.? Of 70°? Of 54°?

54. *From the Equator toward the North Pole*, the temperature diminishes about 1° for every 100 miles.

55. *From the Level of the Ocean* to the summit of a mountain, the temperature diminishes about 1° for every 350 feet.

FROZEN REGIONS.

56. The *Upper Part of this Picture* represents the regions of perpetual snow among the tropical Andes, which correspond, in temperature, to the Frigid Zone.

These *High Snow-clad Peaks* are the great condensers which bring down moisture from the atmosphere, and supply the rains which fill the lakes and rivers of South America.

TEMPERATE REGIONS.

57. The *Middle Portion* of the picture represents a region whose climate corresponds to that of the Temperate Zone.

This *Region* contains plateaus and elevated cities, whose inhabitants enjoy a cool and salubrious climate.

Depressions on the surface of the plateaus form the beds of elevated lakes and streams which receive their waters from the melting snows above them.

Here are Fertile Fields of grain and grass; here flourish trees, fruits, and plants peculiar to the Temperate Zone.

TROPICAL REGIONS.

58. *Below the Line* which marks an elevation of 5,000 feet above the level of the sea, is the climate which corresponds to that of the hot zone of the earth, not only in temperature, but also in its vegetable productions and species of animals.

At various Heights, are deep ravines and fearful precipices, down which rush streams and waterfalls.

FROZEN REGIONS.

59. The *Highest Peaks* of the Tropical Andes are elevated above the level of the sea about 20,000 feet.

The *Most Noted* are Chimborazo, Sorata, Illimani, Antisana, Cotopaxi, and Arequipa.

An immense bird, called the condor, builds its nest far up these heights, and has been known to fly above the summit of Chimborazo.

TEMPERATE REGIONS.

60. The *City of Potosi* is represented on the right of the illustration. It is built on a plateau, at an elevation of more than 12,000 feet above the level of the sea, and contains about 80,000 inhabitants.

Quito is represented on the left, at an elevation of about 10,000 feet; and, although almost immediately under the Equator, its temperature is that of continual spring.

Surrounded by plains and fertile valleys which are enclosed by lofty mountains, Quito is celebrated for the grandeur of its scenery.

TROPICAL REGIONS.

61. *At the Foot* of these mountains the heat is oppressive throughout the year.

The *Trees* of the lower or hot section comprise the palm, tree-fern, banana, and pine-apple.

The *Animals* comprise the tapir, jaguar, cougar, and several tribes of monkeys; besides, parrots, macaws, and other birds which are noted for the brilliant colors of their plumage.

View among the Andes Mountains, showing that different Zones of Temperature pertain to different Elevations.

62. *A Traveler ascending a High Mountain* of the tropical Andes, passes through climates similar to those of the different zones, from the heat of the Equatorial, to the continual frost of the Arctic regions.

63. *At the Base of the Mountain*, or at the ocean level, he endures the oppressive heat of the tropical sun, and observes the luxuriant vegetation, lofty trees, and luscious fruits of the hot zone.

64. *Half-way up the Mountain*, he enjoys the delightful air of the Temperate Zone, with its characteristic varieties of trees, plants, and grains.

65. *Continuing to ascend*, he observes that the mercury in the thermometer is gradually falling, and passes through regions whose temperature admits only of the growth of low evergreens, stunted shrubs, and mosses.

66. *As the Traveler approaches the Top*, he enters the region of perpetual snow, and experiences a climate similar to that of the Esquimau or the Laplander.

Section XVIII.

VEGETATION; ITS GROWTH AND USES.

1. *From Vegetation*, all animal life derives its food, either directly or indirectly. Some animals subsist on flesh, which, however, is the flesh of animals that have fed on vegetation.

2. *For this Reason*, the Creator has covered the greater part of the land with vegetation; for this reason, He made the grass, herbs, and trees, before living creatures were brought into existence.

" He causeth the grass to grow for the cattle, and herb for the service of man."

3. The *Inhabitants of One Climate* require food different from that required by the inhabitants of another climate.

4. *Differences in Temperature*, soil, and degree of moisture on the earth's surface, produce differences in the kinds of plants, and furnish to the various races of mankind and species of animals, the food which is best suited to their wants.

5. The *Inhabitants of the Hot Zone* require food of a light or watery nature; therefore, that region is provided with abundant and luscious fruits, besides rice, millet, and sago.

6. *When you leave the Tropical Regions* and enter a cooler climate, food of a more substantial nature is required.

7. *In the Temperate Zones*, food is obtained mainly from the heavier grains and the flesh of animals.

8. *In the Frigid Zones*, the inhabitants subsist almost entirely on animal food.

9. It is therefore, *according to a Wise Design* that the tropical regions yield the most abundant vegetation.

10. The *Conditions which are most favorable* to the growth of plants, are heat and moisture.

11. *Trees supply Man with Ripe Fruits* and afford shelter during the hot season; some are cut down and sawed into lumber for building purposes and for fuel.

12. *From Plants*, man obtains food for himself and for the animals which are useful to him.

13. The *Most Important Food Plants* are wheat, corn, rice, oats, rye, and potatoes.

14. *Plants derive their Nourishment* from the water which they receive from the soil through their roots, and from the atmosphere through their leaves.

15. *Plants are provided* with cells or tubes through which the water circulates. Those plants which have the largest cells, roots, and leaves, require most water.

16. *Water holds in Solution* various substances that are contained in the soil and are required for the growth of plants; these are, chiefly, carbonic acid, with animal, vegetable, and earthy substances.

17. *Carbonic Acid Gas* is exhaled from the lungs of animals; and, although poisonous to all living creatures, it furnishes the material which enters largely into the formation of trees, vegetables, and flowers.

18. *Herein is the Economy of Nature* plainly manifested: vegetation sustains animal life; animal life and animal substances sustain vegetation. They depend upon each other.

19. *Vegetation not only furnishes Food* for living creatures, but it also extracts from the air that which would be destructive to animal life. It, therefore, is the means of preserving the atmosphere in a pure state for the well-being of the earth's inhabitants.

20. *When the Water which is within a Plant becomes Frozen*, the plant withers, because the water ceases to circulate.

21. *As Snow usually falls before Severe Frost begins*, it keeps the heat of the ground from passing out into the air, and protects the roots of plants and grasses; hence the farmer always welcomes a heavy fall of snow; for the wheat sown in the autumn is protected and nourished by the snowy covering.

" He sendeth forth His commandment upon earth; His word runneth very swiftly. He giveth snow like wool."

22. The *Soil contains Ingredients* necessary to the life of every plant, whether it be the shade or fruit tree, the cotton or tobacco plant, corn, sugar-cane, or potato; and, as the animal body is so constituted as to draw from its food all the elements necessary to the growth of bone and flesh, so the plant draws from water, air, and soil, the different substances required for the growth of wood, leaves, bark, flowers, and fruit.

23. *Besides Soil, Moisture, and Heat*, plants require the light of the sun.

24. The *Light of the Sun assists* in preparing their nourishment, gives them their green color, and causes their leaves and blossoms to open, and their fruit to ripen.

25. The *Grape does not become Fully Ripe in England and Northern France*, because of heavy fogs, which hinder the action of the sun's rays.

26. *All Animals do not eat the same kind of Food*, neither do different plants and trees draw from the soil exactly the same substances.

27. *Each Variety of Plants must be supplied* with the food or elements, adapted to its wants, or it will not flourish.

28. *This is why the Farmer does not sow the same Seed* in the same field every year, and why he manures the soil; for, otherwise, it would soon become exhausted of the elements required specially by the plant which springs from that seed.

29. *Plants thrive* only where the soil allows the roots to spread, and the air and water to penetrate to them; therefore they do not flourish on rock, or in hard, compact clay.

30. *When the Farmer fails to respond to these Laws*, he is soon reminded of his neglect by the appearance of weeds, which seem to call upon him to uproot them; this done, the soil is loosened, and the labor of the industrious husbandman is recompensed by an abundant harvest.

31. *Plants are greatly dependent* upon the moisture and gases contained in the atmosphere.

32. *Some Plants flourish with their Roots* either in the soil, or in water alone, as the hyacinth. The " air plant " grows without either soil or water, the air affording sufficient nutriment for its growth.

Hyacinth.

Seed of a Maple Tree. Seed of the Thistle.

33. *Vegetation is extended by the Winds and Water*, which carry seeds to great distances.

34. *For this Purpose* some seeds are provided with a kind of wing, some with a downy substance, and others with a waterproof covering; but the distribution of the useful plants is accomplished chiefly by man.

35. The *Potato was first found* in Peru, and was afterwards taken from Virginia to England by Sir Walter Raleigh, in 1586. It is now cultivated in nearly every part of the world.

36. *Wheat, Rye, and Oats* came, probably, from the western part of Asia.

37. The *Seeds of some West Indian Plants* have been carried by the Gulf Stream to the western and north-western shores of Europe; while, on the other hand, the vegetation of one region may be kept distinct from that of a neighboring region by intervening mountain ranges, or deserts.

38. *Vegetation prevents* the soil from being washed away and injured by the rains.

39. The *Winds not only supply Moisture to the Plants*, but they also remove it when the quantity is superfluous.

40. *Plants are distributed* with reference to climate. In the *Hot Zone* grow rice, sago, bananas, dates, cocoanuts, and yams; in the *Temperate Zone*, wheat, rye, Indian corn, oats, and potatoes; while the *Polar Regions* are almost destitute of food plants.

41. The *Climate of the Torrid Zone* not only affords the most luxuriant vegetation, but keeps the trees and plants in leaf throughout the year; while, in the other zones, vegetation diminishes with the distance from the Equator, and the leaves fall every year, at the approach of winter.

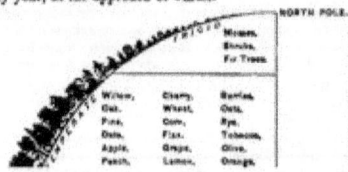

43. *Vegetation in the Northern Hemisphere* extends further north on the western sides of the continents than on the eastern, owing to the agency of the south-west winds which blow over the warm currents of the ocean.

44. The *Forest Trees of the Temperate Zones* are mostly deciduous—that is, their leaves fall in the autumn; some, however, are evergreen, or indeciduous.

45. The *Productions peculiar to the Temperate and Frigid Zones* do not generally thrive in the hot zone, even if transplanted there, unless they are placed in elevated situations where the climate corresponds with that of higher latitudes.

46. *Apples, Pears, and Grapes* belong to the Temperate Zone, and thrive in the Hot Zone only at an elevation of from 600 to 1,000 feet.

47. The *Productions of One Zone* are not separated from those of the adjoining ones by any distinct line, the change from one zone to another being gradual.

48. *From the Base to the Summit of a Lofty Mountain*, vegetation varies with the elevation; on its sides are the same gradations of climate, with their characteristic varieties of plants and trees, that exist on the earth's surface between the latitude of the mountain and the Poles.

49. The *Mountains and Valleys in the State of California* afford every variety of climate, with fruits peculiar to every zone. There flourish the olive, the fig, the date, the grape, the pine-apple, the peach, the apple, and the pear; besides all varieties of grain. In the forests grow mammoth trees, many being from 300 to 400 feet high, and from 25 to 35 feet in diameter.

50. Of what use is vegetation? What kinds of food are adapted to the inhabitants of the Temperate Zone? The Frigid? The Torrid Zone? In what zone do bananas, cocoa nuts, and dates grow? What zone is most favorable to grain, apples, and grapes?

Do different kinds of plants receive their nourishment from the same ingredients of the soil?

In what zone is vegetation the most abundant?

Mention some of the uses of trees and plants. Of snow.

What two elements are necessary to the growth of all plants?

Name the principal trees and plants of the Torrid Zone. Of the Temperate Zone. Of the Frigid Zone. (See illustration on first column.)

What effect have the winds upon the climate and productions of California?

What can you say of the trees of California?

On which side of North America does vegetation extend further north? In what part of the Torrid Zone could you find the climate and productions of the Temperate Zone?

51. *The land which forms the continents* was, at first, but slightly elevated above the surface of the water, and became covered with plants and heavy trees, such as are shown in the

Interior or Sectional View of the Coal Regions of Pennsylvania, showing Strata, which resulted from Successive Submergences of the Surface. The Trees whose Stumps are here represented, flourished at the Earth's Surface in Periods long past.

56. *By Digging downward* in the coal regions, various strata are met with, as shown above; they do not consist of the same materials, nor do they lie in the same order, in all places.

57. *The Distribution of Coal* in various parts of the earth, plainly indicates that its importance to man was anticipated by the Creator. Even the necessity for coal, in the working of iron ore, was provided for by Him; this is observed in the remarkable association of the two.

58. *The Dirt-beds which contain the Roots of Trees and Plants*, formed, at some period, the surface soil which supported vegetation; and the greater the vegetable mass that was submerged, the thicker would be the coal bed; and, while a coal bed extends over considerable space, it is generally much thinner than the strata of sand, clay, and stone, which may be above or below it.

59. *Many Stumps of Large Dimensions*, and with very extended roots, have been found both in America and England, transformed into coal; the stumps retaining their shape and the natural roughness of the bark.

60. The *Vegetation of which Coal was formed*, included the trees and plants of the forests and marshes.

61. *Vegetation which undergoes Decay* on the surface of the earth serves to enrich the soil.

62. *Vegetation which entered into the Formation of Coal* must have been entirely submerged through long periods of time.

63. *Had there been no Submergence* of vegetation, we would not now be provided with coal.

64. The *Different Coal Beds*, lying one below the other, show how often that part of the surface was above the water level, and covered with vegetation.

65. *In Nova Scotia*, there have been discovered nineteen parallel seams of coal, varying in thickness from two inches to four feet.

66. *At the present Rate of Consumption of Coal*, it is estimated that the coal fields of Pennsylvania alone, could meet the demand of the whole world for more than 1,000 years.

Appearance of Parts of the Earth's Surface at the Commencement of the Age of Reptiles. The Fern with other Trees and Plants here represented entered largely into the Formation of Coal.

Section XIX.

ANIMALS; THEIR CREATION AND USES.

1. *Vegetable and Animal Life existed* long before the creation of man, and mutually contributed to each other's support and nourishment; vegetation sustaining animal life, and the decay of animal bodies and substances, through long ages, adding to the fertility of the soil.

2. *Soil that is destitute of Decomposed Animal or Vegetable Substances* is very poor, and will yield little or no vegetation; such was the condition of vegetable life at its commencement; such, also, was the beginning of animal life—very inferior in character and form.

3. *An Improvement in the Quality of the Soil*, caused an improvement also in the varieties of plants; following which came different and improved species of animals.

4. *Geologists show that the Animals which were first created* were very different from those we now see upon the land.

5. *Those first formed* were of the simplest construction, hardly distinguishable from plants.

6. *Different Kinds or Classes of Animals* followed each other; each class being superior in construction, powers, and usefulness, to those which preceded it.

7. *Throughout the Works of Nature*, we see the leading law of development—improvement by successive steps.

8. *According to this Law*, from a small seed springs a tender plant, which enlarges gradually until it becomes a great tree.

9. The *Mighty River* started upon its course as a mere rivulet, which was formed from a trickling spring.

RADIATES.

Jellyfish. Starfish. Actinia. Coral. Medusa. Polype. Actinia.

10. *Animal Life first appeared* in the form of *Radiates.* After them came *Mollusks,* then *Articulates;* after these there followed in order, *Fishes, Reptiles,* and *Mammals.* Last of all came *Man.*

11. *A Knowledge of the Animals which preceded Man* is obtained by digging into stratified rock, where their forms, sizes, and construction are distinctly observed. (*See page 8, paragraph 10.*)

12. *Radiates,* in construction, resemble a flower or plant, but differ from them in having a mouth and stomach. Their bodies are nearly transparent, and seem only to float or rest in water.

MOLLUSKS.

Nautilus. Squid. Scallop. Clam. Oyster. Snails.

13. *Mollusks* are those which have soft bodies without bones or skeletons; some are naked, while others are enclosed in shells for their protection. Of the latter, oysters, clams, and snails furnish examples.

14. *Articulates* are characterized by jointed or articulated coverings consisting of a series of rings: they comprise such animals as worms, crabs, lobsters, spiders, and winged insects.

ARTICULATES

Common Mosquito. Mosquito. Butterfly. Lobster. Beetle. Caterpillar. Grasshopper.

15. *Following the Creation of Articulates was that of Vertebrates,* which embrace all animals having a backbone.

16. *The First Vertebrates* were fishes, then reptiles, birds,

17. *Mammals* are those animals which breathe with lungs, suckle their young, and have warm blood. They include Mankind (*bimana—having two hands*), the Monkey (*quadrumana—having four hands*), and the following named animals:

CARNIVORA, OR FLESH-EATERS			RUMINANTS, OR CUD-CHEWERS		RODENTS, OR GNAWERS	
Lion,	Panther,	Bear,	Ox,	Deer,	Hare,	Beaver,
Tiger,	Dog,	Walrus,	Sheep,	Camel,	Rabbit,	Rat,
Leopard,	Cat,	Seal.	Goat,	Giraffe.	Squirrel,	Mouse.

PACHYDERMS, OR THICK-SKINNED ANIMALS		EDENTATES, OR TOOTHLESS	CETACEA, OR SEA MAMMALS	INSECTIVORA, OR INSECT EATERS
Elephant,	Horse,	Sloth,	Whale,	Mole,
Hippopotamus,	Zebra,	Ant-eater,	Porpoise,	Rat,
Rhinoceros,	Hog.	Armadillo.	Dolphin.	Hedgehog.

18. *Animals of the Different Zones.*

IN THE ARCTIC REGIONS OF BOTH HEMISPHERES.

The Reindeer,	Polar Bear,	Whale,	Seal.

IN THE TEMPERATE ZONES OF BOTH HEMISPHERES.

Horse,	Ox,	Sheep,	Deer,	Wolf.

IN THE TEMPERATE ZONE.

North America,	Grizzly Bear,	Bison,	Puma,	
Europe,	Brown Bear,	Chamois,	Wild Boar,	Stag.
Asia,	Tiger,	Camel,	Musk,	Deer Sable.

IN THE TORRID ZONE.

South America,	Jaguar or American Panther,	Puma,	Tapir,
	Llama,	Alpaca,	Sloth, Monkey.
Asia,	Camel,	Tiger,	Elephant, Rhinoceros,
	Asiatic Lion,	Panther,	Crocodile, Monkey.
Africa,	African Lion, Camel,	Hippopotamus,	Antelope,
	Camel'opard or Giraffe,	Zebra,	Hyena,
	Leopard,	Orang Outang, Ape,	Monkey.

19. It is believed that the *Submergence, at Different Periods,* of vegetation which entered into the coal formations, occurred before the creation of birds; and with vegetation, sank also vast collections of animal bodies, such as mollusks, insects, fishes, and reptiles, which contributed largely to the formation of the strata beneath the present surface of the earth.

20. *The Earth yields Productions and Species of Animals* peculiar to each region or climate.

21. *The Largest Animals* are in the hot regions; they are the elephant and hippopotamus, whose covering is a tough skin, almost entirely destitute of hair; while, in the Arctic regions, where it is too cold for the horse and the ox, live the reindeer and Polar bear, thickly covered with hair, to protect them from the severe cold.

22. *The Near Approach of America to Asia,* at Behring Strait, has given to the Arctic regions of both continents the same species of animals.

23. *The Reindeer and Polar Bear* abound in the Arctic regions of North America, Europe, and Asia.

24. *Animals are adapted to the zones* and districts which they inhabit; their wants and uses are wonderfully fitted to the circumstances in which they are placed.

25. *In the Temperate and Warm Zones* is found the horse,

Laplanders on their Sleds drawn by Reindeers.

26. *In the Frozen Regions of the North*, are found the reindeer and the seal.

27. The *Reindeer* constitutes almost the entire wealth of the Laplander, furnishing him with flesh and milk for food, and drawing his sledge over vast fields of snow.

28. *These Animals obtain their Food* from mosses and low plants, for which they root through the snow, like swine in a pasture.

29. The *Esquimaux derive their Support* from the seal, and exert their greatest energies in the capture of this aquatic mammal.

30. The *Flesh and Fat of the Seal* are used for food; its oil, for light and fuel; the skins are made into clothing, leather, boats, and tents, and form an important article in the fur trade.

31. *Seals are found* in large numbers on fields of floating ice near the coast of Greenland.

32. The *Camel* was made for the desert, where the burning climate and the absence of water render all other animals useless to man.

33. *Providence has given to the Camel* a kind of reservoir or system of cells in which to carry a supply of water sufficient for a long journey; it is also furnished with sharp teeth to cut the few tough shrubs of those barren tracts; and, that it may not be suffocated by the driving sand and dust, its nostrils are so formed as to allow respiration without admitting sand. Its feet are provided with a kind of pad or cushion which prevents their sinking into the soft and yielding sand.

34. *Some Animals inhabit* the dry land, some the water, some fly in the air, and others have the power of living either on land or in water. These last are called amphibious.

35. *A Bird was not formed to live in Water*, like a fish, hence it is not covered with scales; a fish cannot live in the air and find its food among the trees; therefore, it is not provided with feathers and wings; the elephant, the horse, and the ox are unlike both the bird and the fish; but according to their several requirements and uses, they have received their forms, powers, and places.

36. *Animals, like Plants, abound* most in the hot zone, and least in the frigid.

37. The *Surpassing Abundance*, in South America, of vegetation and of the lower species of animals, such as insects and reptiles, is attributable to the excessive heat and moisture of its tropical regions.

Section XX.

MANKIND; THE RACES.

1. "Thus the heavens and the earth were finished, and all the host of them. And the Lord God formed man of the dust of the ground, and breathed into his nostrils the breath of life; and man became a living soul."

2. *For what Purpose was man created?* (See page 5.) Was man created before, or after, animals? Why? Were grass, plants, and trees made before, or after, the creation of animals? Why?

3. *Man is distinguished from all other Animals*, not by his form only, but by his powers of reason and speech. He acknowledges the infinite goodness, wisdom, and power of the Creator, and seeks to advance continually in wisdom and happiness.

4. *Man's Constitution* is such that he is capable of living in any latitude, from the hot to the frozen zone; or at any elevation between the level of the sea and the region of perpetual snow on the sides of mountains.

5. *However Extreme may be the Coldness* of the climate which man enters, his dominion over the animal, vegetable, and mineral kingdoms enables him to procure from them clothing and fuel, which compensate for the lack of solar heat.

6. *While mere Animals are restricted to a Few Varieties of Food*, man partakes of the fruit and vegetables of the soil, and of the flesh of creatures which inhabit the land, the water, and the air.

7. *Mankind is divided into Five General Classes*, or races: the Caucasian, or white race; the Mongolian, or yellow race; the Ethiopian, or black race; the Malay, or brown race; and the American Indian, or red race.

8. The *Races are distinguished from each other* by the color of the skin, kind of hair, and structure of the body and the skull.

9. *These Differences* are produced chiefly by differences in climate, food, and pursuits.

10. The *Influences of these Conditions* upon the physical and mental characteristics of man are vast and unavoidable.

11. *Change the Climate of a Country* either in degree of temperature or of moisture, and a change will be effected also in the character of its vegetation, in the number and kinds of its animals, and in the temperament and pursuits of the inhabitants.

12. The *Condition of a Nation* would be affected by a material change in its systems of rivers, canals, and railroads.

13. *Improved Means of Intercommunication* serve to advance the civilization, education, and prosperity of the people, and to promote a spirit of national unity.

14. *This is obvious in the United States*, where constantly increasing lines of travel by railroads, steamboats, and canals, together with elaborate postal and telegraph systems, contribute largely to the growing power of this republic.

15. The *Depressing Effects of the Absence of these Means* of development are observed in the condition of Africa and the greater part of Asia.

16. *Races and Nations are adapted to the Climate* of whatever portion of the earth they inhabit.

17. The *Hindoo and the Ethiopian* prefer their hot zone, with its light, vegetable food.

18. The *Esquimaux and the Laplanders* cling with strong attachment to their boundless fields of snow, obtaining their subsistence from the animals and fish of the Arctic regions.

19. The *Greenlanders* have their habitation between 70° and 80° north latitude, while the Red Man of South America, and the Blacks of Africa, live under the burning sun of the Equatorial regions.

20. The *White Inhabitants of North America and Europe*, accustomed to a temperate climate, can live in either of these extremes, and on almost every variety of food.

21. *Europe Colonized the Temperate Zone* of North America with wonderful success, but the results of her efforts in other zones have been, comparatively, failures.

22. *In the Tropical Part of Asia*, is British India, which is celebrated for the richness of its productions,—the cotton-plant, sugar-cane, silk, and all varieties of fruits, besides gold, diamonds, precious stones, and nearly all the metallic ores; but, notwithstanding England's influence and authority in that section for more than a century, there is yet only one white inhabitant for every 3,000 natives.

23. *In the Tropical Regions*, the inhabitants subsist, to a great extent, upon the spontaneous yield of the soil; this, together with the enervating influence of the oppressive heat, causes them to lack energy, industry, and patriotism.

24. *In the Frozen Regions*, the inhabitants are dwarfed both in physical stature and mental powers; this is owing to the severity of the climate, with the absence of natural productions and of inducements to labor.

25. *Hardships, Want, and Continual Cold* in the Frigid Zone, and luxury, indulgence, and continual heat in the Torrid, retard the development of their inhabitants.

26. *Both of these Regions* lack that *diversity* of climate and of other conditions, which is necessary to the promotion of individual and national prosperity.

27. *In the Temperate Zones* are enjoyed the greatest blessings which the earth affords. Their lands are neither parched nor icebound; neither teeming with enervating luxury nor stinted to shrubs and mosses; their position on the globe, their systems of mountain ranges, ocean currents, and their change of seasons, combine to promote among the people, that spirit of energy and enterprise essential to their development and happiness.

34. The *Caucasian*, or white race, comprise the most powerful and enlightened nations of the world.

35. *They inhabit* nearly all that part of North America which lies south of the parallel of 50° north latitude, or that part south of the northern boundary of Canada; along the coasts of South America; the greater part of Europe; western and south-western Asia; northern and north-eastern Africa.

36. The *Mongolians*, or yellow race, have thin, coarse, and straight hair, low foreheads, wide and small noses, and thick lips.

37. *They are more numerous* than any other race.

38. The *Mongolians inhabit* the Arctic regions of both continents, and all Asia, except its western and south-western parts.

39. The *Chinese, Japanese, and Esquimaux* belong to the yellow race.

40. The *Ethiopians*, or black race, thrive in the heat and dampness of the tropics, where the white man soon dies.

41. *They Inhabit* nearly all that part of Africa which lies south of the Great Desert.

42. The *Egyptians, Abyssinians, and Berbers*—the inhabitants of Barbary—are Africans, but not Negroes. They belong to the Caucasian race.

43. The *Malays* are of a reddish brown color; their hair is black, straight, coarse, and abundant.

44. The *Malays* are treacherous, ferocious, and less sensible to pain than the other races.

45. *They inhabit* the Malay Peninsula, Sumatra, Java, New Zealand, and many other islands of the Indian and Pacific Oceans.

46. The *American Indians*, so called by Columbus, are copper-colored, tall in stature, and have straight, black hair.

47. Before the arrival in America of the whites, the Western Continent was inhabited by the red men, excepting, however, in the Arctic regions and Greenland, which are inhabited by the Esquimaux.

48. The *Esquimaux* are classed among the Mongolians, in which race many authorities include also the Indians of America.

49. The *American Indians*, in disposition, are melancholy, revengeful, and jealous, and feel bodily pain less acutely than the whites.

50. The *Red Men and the Esquimaux* of America entered that division from Asia, probably in the direction of Behring Strait.

51. The *Human Family had its Origin* in Western Asia, whence it extended into all lands. From the race that moved westward and peopled the lands bordering on the Mediterranean Sea, sprung nations celebrated in ancient history for their progress in civilization and learning.

52. *In Africa*, were ancient Egypt and Carthage; and in Europe, were Greece and the Roman Empire.

53. The *Wave of Progress and Power* continued to roll westward to the Temperate regions of the New World, now the United States of America.

"WESTWARD THE COURSE OF EMPIRE TAKES ITS WAY."

54. *Columbus sailed Westward;* and, by his discovery of the Western Continent, two worlds became acquainted with each other, for their mutual development and advantage. One contributed its vast natural resources; the other, its blessings of civilization and vigor of intellect.

55. The *New World was near enough to the Old* to receive aid while in its infancy, and far enough from it to demand of its new inhabitants the most active employment of their energy and skill toward the development of its resources.

56. The *New World has grown* in usefulness, greatness, and influence with wonderful rapidity.

57. The *North Temperate Zone of America* is vast in vegetable, mineral, and commercial wealth, and contains a people renowned for their energy, enterprise, and achievements, both in peace and in war.

58. *As each Successive Period in the Creation of the Earth* was marked by improvement, so the American Nation is recognized as rising above all others in the sphere of usefulness, development, and influence.

59. The *Productive Plains of the Center and South*, the manufacturing region of the north-east, the broad plains and rich mines of the west, united by easy lines of communication and occupying positions perfectly adapted to each other—plainly show that Providence designed this nation to be ONE and INDIVISIBLE.

NOTE.—The teacher will here turn to the "INDEX TO CONTENTS ARRANGED AS A GENERAL REVIEW OF PHYSICAL GEOGRAPHY," which may be found near the end of the book, and divide it into lessons of convenient length for the class.

INDEX TO CONTENTS

ARRANGED AS A

GENERAL REVIEW OF THE PHYSICAL GEOGRAPHY.

	Page	Paragraph
Africa.—Describe its Plateaus and Mountains	16	45–48
What can you say of *its Inlets?*	13	43
What is the effect of its lack of the means of communication?	44	13–15
What can you say of *its Inhabitants?*	46	42
What were its *Celebrated Nations?*	46	52
Where do some of its *Rivers* empty?	30	41
The *Nile*—Whence is it supplied?	30	52
Alps.—Their *Height*—What is it?	16	45
The *Highest Peak* of the Alps—Mention it	16	45
Their *Passes*—What can you say of them?	18	72
Their *Limit of Perpetual Snow*—At what elevation is it?	16	23
Amazon River.—Its *Source*—Where are they situated?	28	6
Its *Supply*—Whence and how is it received?	29	15
Its *Basin*—What is its area?	29	30
Andes.—Their *Height*—How compared with the Rocky and the Appalachian Chains?	17	54
Their *Slopes*—Describe them	17	55
Their *Influence* upon Rain and Climate—What is it?	17	55–59
Their *Position*—What can you say of it?	17	62
Animals.—Were all Species created at once?	7	12
Those *first formed*—What was their character?	42	5
Their *Development*—What can you say of it?	42	6
What *General Name* has been given to those first formed?	43	10
Radiates—Describe them	43	12
Name some of them	43	Cut.
What Species succeeded *Radiates?*	43	10
Mollusks—Describe them	43	13
Name some of them	43	Cut.
What Species are third in the order of Creation?	43	10
Articulates—Describe them	43	14
Name some of them	43	Cut.
Animals.—Mention those in the Torrid Zone of South America	43	18
Mention those in the Torrid Zone of Asia	43	18
" " " " Africa	43	18
Are Animals adapted to Climate?	43	21, 24
The *Reindeer*—What can you say of it?	44	27, 28
The *Seal*—What can you say of it?	44	29–31
The *Camel*—What can you say of it?	44	32, 33
In what Zone are Animals most numerous?	44	36, 37
How much of an Animal Body consists of Water?	22	2
Upon what do Animals subsist?	40	1
How do Animals and Plants mutually depend on each other?	40	17, 18
Antarctic Current.—Describe it	25	54
Arctic Currents.—Describe them	25	23
What do they deposit off the Coast of Newfoundland?	25	24
Their *Influence*—How felt upon the East Coast of the United States	37	24
Artesian Wells.—Their *Formation*—Explain it	28	26–28
Their *Name*—From what derived?	28	26–28
Their *Depth*—to what Depths have come been sunk?	28	30–34
Temperature—Whence is it derived?	28	30
Describe an Artesian Well at St. Louis—At Charleston	28	33, 34
Asia.—Its *Surface*—Describe it	16	44
What is the Mean Elevation of the Land?	16	85
River Systems—What can you say of them?	30	30
How do they compare with those of Europe?	30	57
Its *Area*—What is it?	14	53
Its *Highest Point*—Mention its name and height	16	85
Atlantic Ocean.—Its *Area*—How many Square Miles	22	13
Greatest Depth—Where?	23	21

	Page	Paragraph
Atmosphere.—Its Temperature—How derived?...	32	12
Is the Upper or the Lower Part the warmer?...	32	12
How is its Temperature regulated?...	26	41
Its Movements—Mention them...	32	16
Its Capacity of holding Water—How increased and diminished?...	33	5
Its Uses—What to plants?...	40	14
How influenced by Vegetation?...	40	19
Boulders.—Describe their Origin and Formation.	15	14
Caucasians.—What people do they comprise?...	46	34
They Inhabit—What part of North America? South America? Europe? Asia? Africa?	46	35
Chinese.—To what race do they belong?...	46	39
Cities.—Mention the most elevated in the World?	17	Cut.
Their Location—Inland, or near navigable Waters?...	30	58
Climate.—What is Climate?...	26	1
Upon what does it depend?...	23	26
It is modified by what?...	9	11
In what parts of the Earth is it most uniform?...	33	33
Why is the Land warmer than the Water, in Summer?...	33	13
Why is the Land cooler than the Water, in Winter?...	33	14
Which is the warmer Side of the Eastern Continent—the Eastern or the Western? Why?	36	7, 8
Traveling Eastwardly from the Atlantic Coast of Europe what change of Temperature is experienced? Why?	36	9
Which possesses the warmer Climate—France or Newfoundland? Why?...	36	10
Which has the more uniform Climate—The British Isles or Labrador?...	36	Chart.
Is the European or the American Side of the Atlantic the warmer?...	36	11, 12
Into how many and what Climatic Zones is the Northern Hemisphere divided?...	37	15
Between what Lines are Climatic Zones included?	36	6
What is the Mean Annual Temperature of the Frigid Zone? The Cold Zone? The Temperate Zone? The Warm Zone? The Hot Zone? The Torrid Zone?...	37	16
What can you say of the Climates of the Western Coasts of the United States and Europe?...	37	21
What is the Mean Temperature of the Hottest Month in New York? In San Francisco? Of the Coldest Month in New York? In San Francisco?	37	22
What is the Mean Difference in Temperature between Summer and Winter, in New York? In San Francisco?	37	22
In which of these two Cities is the Climate excessive? Uniform?	37	22
What amount of Snow falls in New York? In San Francisco?	37	23
Of what does the Winter of San Francisco mostly consist? The Summer?...	37	23
Climate.—What Ocean Currents reduce the Temperature of the Atlantic Coast of the United States?	37	24
What is the Climate of the Valleys in Western California?...	37	25
Compare the Climates of the Faroe Islands with that of Yakoutsk?...	37	26
In which is the Climate excessive? Uniform? Why?...	37	26
What is the Climate of the Azores and Madeira Islands?...	37	28
What Cities of the United States lie between the same Parallels as these Islands?...	37	28
Which Side of North America possesses the Warmer and more even Climate?...	37	38
On what part of the Earth's Surface is the Climate most uniform?...	37	32
Compare the Climate of Vancouver's Island with that of Maine?...	37	33
As you leave the Equator and approach the Poles, what changes of Climate are experienced?...	26	52
What Climates are experienced on the Sides of Tropical Mountains?...	30	53
What is the Mean Temperature at the Equator?	38	Chart.
At the Foot of a Tropical Mountain?...	38	Chart.
At 30° North Latitude?...	38	Chart.
At what Elevation would the Temperature be 70°?	33	Chart.
What part of a Tropical Mountain represents the Climate of Greenland? Of the United States? Of the Torrid Zone?...	38	Chart.
At what rate does the Temperature diminish between the Equator and the Poles?...	38	54
At what rate does the Temperature diminish between the Level of the Ocean and the Summit of a Tropical Mountain?...	38	55
Clouds.—What are they?...	34	18
How many and what Classes of Clouds are there?	34	21-25
Describe the Cirrus—The Stratus—The Cumulus—The Nimbus...	34	21-25
How are Clouds Influenced by High Mountains?	17	57
" " " by Winds?...	17	57
How far above the Earth's Surface do Clouds rise?	34	18
Coal.—Its Formation—Describe it?...	41	50-52
Describe the Strata of some Coal Regions...	42	56
What have been found in these Strata?...	42	59
What can you say of the Quantity of Coal known to be in the Earth?...	42	66
Describe the principal Coal Fields of N. America.	42	54
" " " of the Eastern Continent	42	55
Continents.—Their Formation—Describe it...	9	6
Their Number and Names—Mention them...	10	1
The Direction of the Eastern? Of the Western?	12	22, 23
The Form of the Continents and their Divisions?	12	29
Crust of the Earth.—Its Formation—Describe it.	8	1
Of what is it composed?...	8	2
Its Thickness—What can you say of it?...	8	5-7
Its Greatest Depressions—Where are they?...	9	10

	Page	Paragraph
Currents of the Ocean.—*Their Theory*—Explain it.	23	6–9
Illustrate the Movement of the Equatorial Current by means of a Boat Race.	24	11, 12
Their Change of Direction—How caused?	24	8
What gives the Gulf Stream a *Rotary Motion*?	24	Cut.
If South America had not been raised from the Bed of the Sea, what would be the Direction of the Equatorial Current?	24	18
Equatorial Currents of the Pacific—Describe them.	24	19, 20
Cold Currents—How many and what are they?	25	Chart.
Warm Currents—Mention them.	25	Chart.
What Current washes the Eastern Coast of the United States.	25	23
What Current washes the Western Coast of Europe?	25	30
Benefits of the Oceanic Currents—What are they?	25	27
Dead Sea.—*Its Origin*—Describe it.	31	14
What is its Distance below the Level of the Sea?	31	14
What *Substances* are contained in its Waters?	31	14
Deserts.—What are they?	21	1
By what are they *Caused*?	21	2
The Desert Region of the *Old World* comprises what?	21	8
What is its *Extent*?	21	3
Simoon—Describe it.	21	5
Drifting Sand—What destructive Effects?	21	7
Sahara—State its Extent and Elevation.	21	10
Oases—Describe them.	21	11, 12
Atacama—Describe this Desert.	21	14
Dew.—*Its Formation*—Describe it.	33	5
What are its *Uses*?	33	5
Earth.—*Its Creation*—What was the Process?	6	1
Illustrate its Formation from Chaos.	6	2
For what Purpose was it Created? By whom?	5	3
General Order of Creation—Mention it.	7	18
Its Shape—What is it?	6	3
Its Surface—Of what did it at first consist?	6	3
Earthquakes.—*Their Origin*—Describe it.	19	1
Their Effects—Mention some of them.	19	7
How are they rendered less Destructive?	19	5
What *Warnings* precede them?	19	15
What can you say of the Destruction of *Herculaneum and Pompeii*?	19	16
Europe.—*Its Mean Elevation*—What is it?	18	85
Its Great Plain—What Countries are comprised in it?	21	14
Is any part of its Surface below the Sea Level?	21	17
Describe the Region around the Caspian Sea.	21	18
What is the Character of the Land toward the Arctic Ocean?	21	19
What can you say of the *River System* of Central Europe?	20	35
Fissures.—What are they?	15	13
Their Origin.	19	4
Fog.—What is Fog?	34	19
How are the Fogs near Newfoundland formed?	25	26
Food.—From what is it obtained?	40	1
Do all People require the same kind of Food?	40	3
Is it *Adapted* to the Wants of the Earth's Inhabitants?	40	4
What kind is required in the *Hot Zone*? In the *Temperate Zone*? *Frigid Zone*?	40	5, 6, 8
What forms the Chief Food of the *Esquimaux*?	44	29, 30
Geysers.—*Their Position and Origin*?	27	20
How can you illustrate them?	27	20
Give an account of the Eruptions of the *Great Geyser*.	27	22
Glaciers.—Describe them.	16	39
Gulf Stream.—Whence does it proceed? Describe it.	24	9
How is Europe benefited by the Gulf Stream?	24	16, 17
Does it wash the Eastern Coast of the United States? Why not?	25	23
What is the Difference in *Temperature* between the Gulf Stream and the Cold Current near the Coast of the United States?	25	28
Does any part enter the Arctic Ocean? How?	25	Chart.
What is the *Velocity* of the Gulf Stream?	25	31
How far North is its Influence felt?	36	14
How has it assisted in the *Extension of Vegetation*?	41	37
Hail.—How is it produced?	34	31
Heat.—Does the *Internal Heat* of the Earth extend to the Surface?	8	8
Whence does the Surface receive its *Heat*?	8	2

	Page	Paragraph
Lakes.—What are they?	31	1
How many Classes of Lakes are there?	31	2
Describe the First Class—The Second—The Third—The Fourth	31	3-7
How are they supplied?	31	9-10
Why do not all Depressions contain Lakes?	31	8
Mention the Most Elevated Lake and its Elevation.	17	Cut.
What is the elevation of Lake Titicaca?	17	52
What Lake is furthest below the Sea Level?	31	14
Why is the Water of some Lakes Salt?	31	15
Mention the principal Salt Lakes	31	16
Subterranean Lakes—What are they?	31	18, 19
What are sometimes caused by them?	31	18, 19
What is the Largest Lake in the World? Its Area?	31	21
Land Slides.—Describe them	27	23
Man.—How distinguished?	44	3
Is he influenced by Climate?	44	9-11
His Adaptability to Climate—what can you say of it?	44	4-5
The Races—Mention them	44	7
How are they distinguished from each other?	44	8
What can you say of the Races which inhabit the Torrid and Frigid Zones?	45	16-19
How is Man affected by extreme Heat? Extreme Cold?	45	25
What are the Characteristics of the Inhabitants of the Tropical Regions?	45	23
Of those of the Frigid Zones?	45	24
Describe the Inhabitants of the Temperate Zone.	45	27, 33
Where has Man reached the highest State of development?	45	30
Mississippi River.—Describe it	29	19
From New Orleans, how far North is it navigable?	29	20
" " " " North-east? North-west?	29	21, 22
Its Windings—What can you say of them?	29	25
Its Basin—What is its Area?	29	30
Its Delta—How is it formed?	30	44-46
Its Wearing and Transportation Power—How shown?	30	45
Mountains.—Their Origin?	17	64
Time occupied in their Formation?	15	15
A Chain—A Culminating Point—What are they?.	14	5
On the Eastern Continent—Mention them	12	25
On the Western Continent	12	26
Ocean.—How divided? Total extent of Surface?..	22	13
What is the Area of the Pacific? Atlantic? Indian? Arctic? Antarctic?	22	13
Its Bed—What Changes has it undergone?	6	10
Its Temperature—How regulated?	26	89
Its Purity—How preserved?	22	6
By what Process is the Land supplied with Water?	22	4
What are Dependent upon the Ocean?	22	3
Its Uses—Mention the principal	26	43
As a Means of Communication, which Ocean is the most Useful to Man?	22	15
Its Depth—Is it uniform?	22	17
Where is the deepest part of the Ocean?	23	21
What is its Mean Depth?	23	23
What can you say of its Depth near the Coasts?.	22	18
Plains.—What are they?	14	8
What do they comprise?	20	5
How shaped and fertilized?	11	13
Of North America—What do they comprise?	20	7
Of South America—What do they comprise?	21	8
Of the Amazon—What is their Extent?	21	11
The Arctic Plains—Describe them	21	19
Plants.—Their Growth—How does it progress?	6	1
Of what Element are Plants chiefly composed?	22	1
What Conditions are most favorable to its growth?	40	10
Their Nourishment—From what received, and by what means?	40	14, 15
To what do Plants supply Nourishment?	40	12
How are they affected by Frost?	40	20
By what means is the Growth of some Plants extended?	41	34
Mention the Trees and Plants of the Frigid Zone —Of the Temperate Zone—Of the Torrid Zone.	41	41
Plateaus.—What are they?	15	11
Their Formation—Describe it	15	13
Where are the Plateaus of Asia? Europe? America? Africa?	15	24
The Highest on the Globe—Mention them	16	41
The North American Plateau—Describe it	18	77
Rain.—How produced?	34	30
Its Distribution over the Surface—How caused?	34	36
Its Uses—What are they?	26	4
How does it penetrate the Ground?	26	7
How influenced by high Mountains?	34	28, 29
Where does the greatest Amount fall? Why?	34	32

Rainless Regions.—They include what parts of the Eastern Continent?.............. 21 | 3
What parts of the Western Continent...... 35 | 60

Rivers.—How Formed?........................ 28 | 1
Their Uses—Mention them.............. 28 | 7
Their Courses—Describe them—What do they indicate?.................. 28 | 8
The Ganges—Describe it.............. 28 | 13
The Indus and Brahmaputra—Describe them.... 29 | 14
How do Rivers affect the Surface of Lowlands?.. 29 | 18
Deltas—How formed?................. 30 | 14
Their Windings—What advantages attend them?.. 29 | 24
What is the most important River in N. America?. 29 | 19
A River Basin—What is it?............ 29 | 29
A River Bed—What is it? A Channel?....... 29 | 33
A River System—What is it?........... 29 | 28
Island Banks—Name some of them...... 30 | 40
Where do some Rivers of Africa empty?...... 30 | 41
Oceanic Rivers—What are they?........ 30 | 42
Continental Rivers—What are they?...... 30 | 43
How are Rivers affected by the Melting of Snow? 30 | 51
Mountain Streams—Mention their Uses... 30 | 45
Rivers which rise periodically—How supplied?... 30 | 52

Rocks.—What are Aqueous Rocks? Stratified Rocks? 8 | 10
What are Igneous Rocks? Unstratified?...... 8 | 11
Stratified—Of what composed?.............. 8 | 19

Rocky Mountain System.—Its Extent?...... 18 | 79
What Ranges does it include?.......... 18 | 78
The Greatest Width of the System—where and what?............................. 18 | 80

Sea Shells.—To what Class of Animals do they belong?................ 43 | 13
Their Appearance on Mountains—How accounted for?.................... 15 | 14

Snow.—How is it produced?.............. 34 | 32
Of what Advantage is Snow? Why?........... 34 | 35
Of what Use is Snow which covers the Tops of Tropical Mountains?.............. 17 | 58
Perpetual Snow—At what Elevation on the Andes? On the Alps? In Arctic Latitudes?.. 16 | 38

South America.—Its Area in Square Miles?... 14 | 53
Its Plateaus—Where situated?............. 15 | 30

Springs.—Mineral Waters—What are they?...... 27 | 15
Mineral Waters—Of what Uses are they?.... 27 | 15
Mineral Springs—Where are the most celebrated? 27 | 16
Hot Springs—Their Origin and Uses?......... 27 | 13
"   " —Where are the most noted?..... 27 | 19

United States.—Was all the Land of this Country raised at the same time?.......... 11 | 11
Describe its great Plateau Region.......... 18 | 77, 81
For what Production is the Southern Part of this Country noted?............ 38 | 49
What can you say of the North-eastern Part?... 38 | 50
The Means of Communication—What can you say of them?.................. 44 | 14

Vapor.—The Process of its Formation—What is it?. 9 | 16
Its Use—Mention them................ 9 | 16
Why is it not always Visible?............ 33 | 7
To what does it supply Nourishment?......... 33 | 6

Vegetation.—When and for what Purpose was it made?.............................. 40 | 1, 2
The First Vegetation—What was its character?... 42 | 2
Where is it produced in the greatest abundance? 40 | 9
How does it purify the Atmosphere?........ 40 | 19
What Mutual Dependance between Vegetable and Animal Life?............ 40 | 17, 18
On which Side of a Continent does Vegetation extend farthest North? Why?.......... 41 | 43
How does Vegetation vary on the Sides of Mountains?.................. 41 | 46

Volcanoes.—Their Origin?.............. 19 | 1-2
Illustrate by means of a Cake........... 19 | Cut.
Of what Benefit are Volcanoes?.......... 19 | 5
The Effect of an Eruption of Mt. Vesuvius—Give an Account of it............ 20 | 23
Monte Nuovo—Give an account of its Formation. 19 | 10
The Most Noted Volcanoes—Mention them.... 19 | 11
Hot Water and Steam of Volcanoes—Whence do they proceed?.................. 21 | 12

Water.—Whence is the Land supplied with Fresh Water?.................. 9 | 13
The Center of the Water Hemisphere—Where is it?.................. 11 | 18
Its Wearing Power—What can you say of it?... 29 | 26, 27
How affected by Heat?................. 23 | 1
Of what Benefit is it to Plants?.......... 40 | 14

ASTRONOMICAL GEOGRAPHY.

[THE WORDS IN BLACK TYPE SHOULD BE QUESTIONS.]

1. *Astronomical Geography treats* of the form, size, and motions of the earth; its relations to the Sun, Moon, and other heavenly bodies; its seasons, latitudes, and longitudes.

2. *The Earth* is one of a family of heavenly bodies which revolve around the Sun.

3. *The bodies which revolve around the Sun are distributed* into three classes; Planets, Asteroids, and Comets.

4. *These bodies, together with the Sun, constitute the Solar System.*

5. The *Solar System* is but a small portion of the Universe.

6. *The Sun* is a luminous body, because it shines by its own light. The planets are opaque (dark) bodies.

7. *The Earth, Moon, and other planets receive from the* Sun light and heat.

8. *The names of the principal planets,* according to their size, are Jupiter, Saturn, Neptune, Ura'nus, the Earth, Venus, Mars, and Mercury.

9. *Their names according to their distances* from the Sun, are Mercury, Venus, the Earth, Mars, Jupiter, Saturn, Uranus, and Neptune.

10. *The form of the Earth* is that of a sphere, slightly flattened at the Poles. (*See illustration on page 9.*)

11. *A Sphere or Globe* is a round body whose surface, in every part, is equally distant from its center.

12. *A Hemisphere* is half a sphere or globe.

13. *The Diameter of a Sphere* is a straight line passing through its center, and terminated at both ends by the surface.

14. *The Diameter of the Earth* is nearly 8,000 miles.

Its diameter at the Equator is 7,925 miles, but from Pole to Pole it is 26 miles less.

15. *The Circumference of a Sphere* is the distance around it.

16. *The Circumference of the Earth* is nearly 25,000 miles.

17. *The Axis of a Sphere* is the line or diameter on which the sphere revolves.

18. *The Poles of the Earth,* or of any sphere, are the extremities of its axis, or the two points where the axis meets the surface.

19. *The Sun shines upon* one half of the earth's surface at any one time; so that one hemisphere has day while the opposite hemisphere has night.

20. *The succession of Day and Night* is caused by the revolution of the earth on its axis, which it performs every 24 hours.

21. *The rate of Motion* of the equatorial parts is 1,000 miles every hour, but it diminishes toward the Poles.

The Axis does not revolve, neither do the Poles.

22. *Localities on the Earth's surface are determined* and described by means of imaginary lines or circles.

23. *Great Circles* are those which divide the Earth into two equal parts.

24. *Small Circles* are those which divide the Earth into two unequal parts.

25. *The principal Great Circles* are the Equator, Ecliptic, and Meridians.

26. *The principal Small Circles* are the two Tropics and the two Polar Circles.

27. *The Equator divides the Earth into* Northern and Southern Hemispheres. It is midway between the Poles.

28. *Meridians pass* from Pole to Pole, crossing the Equator at right angles.

29. *Meridians divide the Earth into* Eastern and Western Hemispheres.

30. *Latitude* is distance northward or southward from the Equator, measured on a Meridian.

31. *Longitude* is distance eastward or westward from a certain Meridian, measured on the Equator.

32. *Latitude and Longitude are reckoned* in degrees, minutes, and seconds, which are known by the signs (°), ('), (").

The City Hall of New York is in lat. 40° 42' 43" (read 40 degrees, 42 minutes, and 43 seconds). A degree contains 60 minutes, and a minute 60 seconds.

33. *A Degree* is one 360th part of a circle; it varies in length according to the size of the circle.

34. *The length of a degree* on a Great Circle of the Earth is about 69 statute miles, or 60 geographical miles.

A statute mile contains 5,280 feet, and a geographical mile, 6,075 feet.

35. *The parts of the Earth farthest from the Equator are* the Poles, whose latitude is 90°.

36. *Longitude is usually reckoned,* on our maps and globes, from the Meridian of Greenwich, near London, and from the Meridian of Washington.

37. *The greatest Longitude* a place can have is 180°—half way round the globe.

38. *Refer to the Map* on pages 52 and 53, and state the Latitude of Philadelphia; of New Orleans; of Columbus; of Nashville; of San Francisco; of Savannah.

39. *What is the Longitude* of each, from Greenwich, and from Washington?

40. *Refer to the Map* on page 72, and state the Latitude of Naples; of Venice; of Lucerne; of Athens; of Constantinople; of Paris; of Frankfort; of Hamburg; of London; of Liverpool; of Dublin.

41. *What is the Longitude* of London? of Dublin? of Geneva? of Rome? of Vienna?

42. *The Ecliptic* is the path in which the Earth revolves around the Sun. In Geography, the Ecliptic is a great circle on the terrestrial globe which is always in the plane of the earth's orbit.

43. *The Equator and Ecliptic cross* each other at an angle of 23½°.

44. *The Sensible Horizon* is the Small Circle which bounds our view of the Earth's surface. Its circumference is the line in which the Earth and Skies appear to meet; spectators in different localities have different horizons. In the middle of the horizon is the spectator. The higher the elevation on which the spectator stands, the greater is the sensible horizon. A person at sea, standing on the level of the surface, would see three miles in every direction. The diameter of his sensible horizon would be six miles. (*See page 9, illustration, and paragraphs 1 to 5.*)

45. *The Rational Horizon* is the Great Circle which is parallel to the Sensible Horizon; it divides the Earth into upper and lower hemispheres.

46. *Parallels of Latitude* are small circles parallel to the Equator.

47. *The Tropics* are those parallels which pass through the two points of the Ecliptic farthest from the Equator.

48. *The Tropic in the Northern Hemisphere is called the Tropic of Cancer. That in the Southern Hemisphere, the Tropic of Capricorn.*

49. *The Distance of the Tropics from the Equator is 23°.*

50. *The Axis of the Earth is not perpendicular to the plane of the Earth's orbit.*

SMALL CIRCLES

51. *The Distance from the Poles to the Extremities of a Diameter* which is perpendicular to the Ecliptic is 23½°; through those two extremities two parallels of latitude are drawn; that around the North Pole is called the Arctic Circle or North Polar Circle, and that around the South Pole, the Antarctic, or South Polar, Circle. (*See illustration at the top of the page.*)

52. *The Tropics and Polar Circles divide the Earth's surface into five great Belts or Zones.* (*See map on page 51.*)

53. THE ZONES AND THEIR EXTENT FROM NORTH TO SOUTH.

North Frigid....	From the North Pole to the Arctic Circle.............	23½°
North Temperate	Arctic Circle to the Tropic of Cancer........	43°
Torrid.........	Tropic of Cancer to the Tropic of Capricorn...	47°
South Temperate	Tropic of Capricorn to the Antarctic Circle..	43°
South Frigid...	Antarctic Circle to the South Pole...........	23½°
	Total, from Pole to Pole...................	180°

54. *Within the Torrid Zone the Heat* is extreme, because the Sun's rays fall directly upon the surface.

55. *The Cold of our Winter* is not known, except at high elevations. (See page 53, paragraphs 62–66.)

56. *The Days and Nights* on and near the Equator are equal throughout the year. Leaving the Equator, *their inequality increases* with the latitude.

ZONES

57. *The Sun is Vertical to the inhabitants of the Torrid Zone* at certain times during the year. (Read page 45, par. 16, 17, 22, and 23.)

58. *The Sun is Vertical, or in the Zenith, when* it is perpendicularly over the head.

59. *Within the Frigid Zones the Cold* is extreme, because the Sun's rays fall very obliquely upon the surface.

The Longest Days in Summer and the *Longest Nights* in Winter are in proportion to the latitudes,—from 24 hours on the Polar Circles to 6 months at the Poles.

The Sun is never Vertical to any of the inhabitants of the Frigid Zones.

60. *Within the Temperate Zones the Heat* is less than that in the Torrid Zone, and the cold is less than that in the Frigid Zone.

The Longest Days in Summer and the *Longest Nights* in Winter vary from 13½ hours on the Tropics to 24 hours on the Polar Circles.

The Sun is Vertical once a year—midsummer—to the inhabitants on the Tropics.

61. *The Change of Seasons depends* upon the annual revolution of the earth in the same plane, the inclination of its axis, and the leaning of the axis always in the same direction.

62. *The North Pole leans toward the Sun* in the latter part of June; then it is Summer in the Northern and Winter in the Southern Hemisphere. (*See illustration above.*)

The Northern Hemisphere has long days and short nights, while the *Southern Hemisphere has* short days and long nights.

The Whole of the North Frigid Zone has day, while the South Frigid has night.

The Sun is Vertical to the inhabitants on the Tropic of Cancer.

63. *The North Pole leans from the Sun,* in the latter part of December; then it is Summer in the Southern and Winter in the Northern Hemisphere; the *Southern Hemisphere has* long days and short nights, while the *Northern has* short days and long nights.

The Whole of the South Frigid Zone has day, while the North Frigid has night.

The Sun is Vertical to the inhabitants on the Tropic of Capricorn.

64. *On the 23d of March,* neither the North nor the South Pole leans toward the Sun. (*In the illustration above, the pupil must imagine the Earth to have moved around behind the Sun.*) Then it is Spring in the Northern Hemisphere while it is Autumn in the Southern; *the Sun is vertical* to the inhabitants on and near the Equator, and the line of separation between the dark and the illuminated side of the Earth passes through the Poles.

65. *On the 21st of June,* the position of the Earth is as represented in the picture; three months afterward, or on the 23d of September, the Earth's position would be sidewise, as in March. (*In the picture imagine the Earth to have moved toward you, and to be immediately in front of the Sun, about two inches from the page.*)

66. *On the 23d of September* it is Autumn in the Northern, and Spring in the Southern Hemisphere.—12 hours day and 12 hours night, in all the Zones; the Sun vertical to the inhabitants on the Equator; the days and nights are everywhere equal.

HEIGHTS OF MOUNTAINS.

ASIA.

SOUTH AMERICA.

NORTH AMERICA.

EUROPE.

AFRICA.

OCEANICA.

HEIGHTS OF SOME INHABITED SITES.

DISTANCES AT WHICH MOUNTAINS HAVE BEEN SEEN.

RIVERS OF THE WORLD.

NORTH AMERICA.

SOUTH AMERICA.

EUROPE.

ASIA.

AFRICA.

COUNTIES OF WASHINGTON TERRITORY.

BULLION PRODUCT OF 1869.

COMPARATIVE TEMPERATURE
ON THE ATLANTIC COASTS.

www.ingramcontent.com/pod-product-compliance
Lightning Source LLC
Chambersburg PA
CBHW022024190326
41519CB00010B/1590